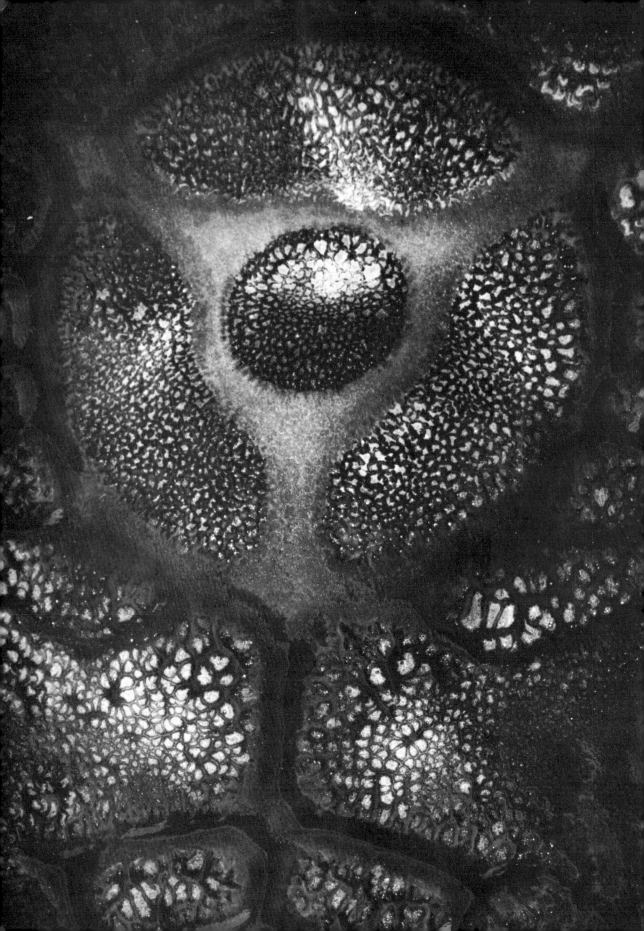

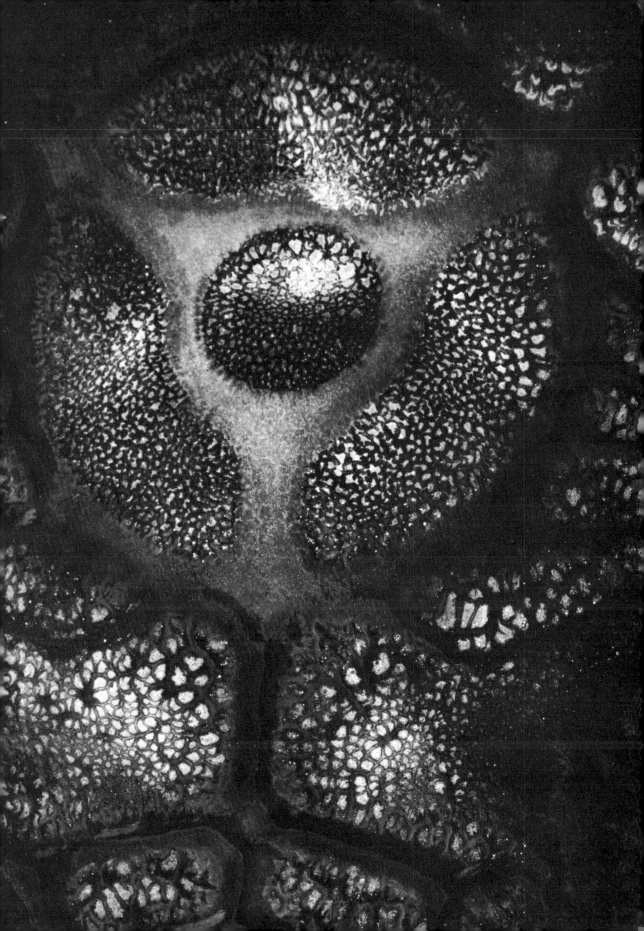

# The Technique of Enamelling

# THE TECHNIQUE OF ENAMELLING

Revised Edition

**Geoffrey Clarke**
**Francis & Ida Feher**

B T Batsford Ltd   London

First published 1967
Reprinted 1969
Reprinted 1972
Revised edition 1977
First published in paperback 1988

© Geoffrey Clarke, Francis & Ida Feher
1967 and 1977

ISBN 0 7134 5875 5

Printed and bound in Great Britain by
Anchor Brendon Ltd
Tiptree, Essex
for the publishers
B T BATSFORD Ltd
4 Fitzhardinge Street
London W1H 0AH

# Contents

# Acknowledgment

*All the black and white photographs, with the exception of those on pages 97, 99 and 100 and the photographs for Chapter III, Section 6, are by Francis Feher, who also did the line drawings.*

The Authors wish to thank the Vitreous Enamel Development Council and the Copper Development Association for the facilities and assistance they so willingly gave, and the British Museum for permitting the enamels on Plate 3 to be photographed. They are also grateful to Mrs Marit Aschan for permission to reproduce her enamel *Sheherazade* on Plate 2, and to Mr W E Benton for the enamels on pages 99 and 100. Mr Benton's kind help in other directions was much appreciated.

The Authors have had the assistance of Valerie Bexley, who has provided the new Chapter III, Section 6, *Control of Design*. Miss Bexley is an enamelling craftsman. The photographs for this Section are by William Eales.

GC and F & IF

# Introduction

Enamelling is the art of fusing glass to metal. It is one of the oldest forms of decoration used.

There is no doubt that enamelling has a special fascination of its own, and those who follow it as a craft, or learn about it as students or collectors, soon fall under its spell. It is certainly unlike any other craft, and perhaps what captures the imagination of the newcomer, and holds his interest for years to come, is the nature of the process itself. The essence of enamelling is the use of heat to marry two dissimilar materials, glass, for that is what enamel is, and metal. In its molten state the enamel fuses to the metal and a lasting bond is created between them. United, they complement each other and form a decorative whole.

Although this book is chiefly concerned with the enamelling of copper, other metals can be enamelled equally well. The main requirement is that the melting temperature of the metal is above that of the enamel. Copper is chosen by most people as it is the cheapest, and, due to its softness, is easy to cut and shape. Silver is second in popularity; although more expensive, it is used quite widely as it enhances the brilliance of transparent enamels to a far greater effect than copper. Gold too can be enamelled: the richness of the combination of this metal and glass is incomparable but it is usually beyond the reach of those taking up a hobby.

Mild steel is being used increasingly in the creation of large plaques and murals. It has to be covered first with a base of white steel enamel before 'jewellery enamels'—those used for the artistic decoration of metals—can be applied. Enamels applied over this white base take on a slightly different appearance from those applied direct to the metal surface as with copper, silver and gold, but the effect is equally beautiful. The advantage of using steel is that it forms a more solid base than copper; it is a far harder metal which does not warp so much, and the enamels when applied to it are not so liable to crack when cooling. However, as it is so hard it is difficult to manipulate, and is thus used for flat plaques rather than three-dimensional or jewellery work (*180*).

Enamelling is not difficult, nor is the equipment out of reach of the ordinary person. The basic requirements of the enameller are metal—in our case copper—enamels, and a method of applying heat. All are quite easily obtained. Equipment may, of course, vary from the simplest and cheapest to the most sophisticated and expensive.

The aim of the authors has been to give guidance to those who have had no experience of enamelling. No special training is required; anyone can take it up as a hobby.

Enamelling is above all creative work, a craft for the individualist who wishes to create the unique. It provides ample opportunity for self-expression, and can be deeply satisfying. The metal may be formed in a variety of shapes and the range of colours of the enamel is virtually unlimited.

Although this book describes only techniques used by the authors, anyone who enamels, while they must follow certain basic rules, can develop their own techniques and methods.

## Enamels

Enamels, which are purchased ready-made, are manufactured by melting together silica, felspar, soda, borax and other materials. The resulting mixture, when in a molten condition, is quenched in water, which shatters the glass, or poured on to a thick slab of steel and allowed to solidify. Colours are provided by metal oxides and other chemicals. Enamel is a wholly inorganic substance, made of materials which come from the earth or the rocks; it has the agelessness and the permanence of the rocks themselves.

Enamel is widely used as a finish on many utilitarian articles in the home, either applied to steel as with a cooker, or cast iron as with a bath. It should not be confused with stove enamel, which is a paint that has been baked on. The word vitreous, which means of glass, is often used to qualify enamel in the commercial world.

But, while enamels have been employed in industry for a mere hundred years or so, they have ornamented and embellished metal since before the birth of Christ. Here is a superb form of decoration, giving a choice of colours ranging from the most dense opaque shades to clearest transparent.

A warning before you start: enamelling does, as we have already said, exercise a strange fascination, and, once having started, you may never be able to give it up.

# How to Enamel

*1* Polishing the back of the bare metal work-piece

*2* Spraying on gum solution

*5* Polishing the face

*6* Applying enamel to the face

Dusting on powdered enamel

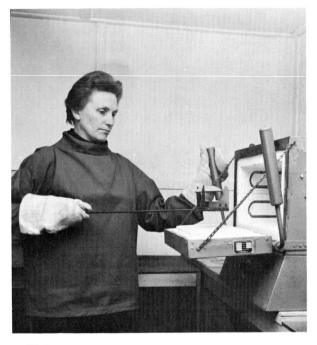

*4* Firing

Firing

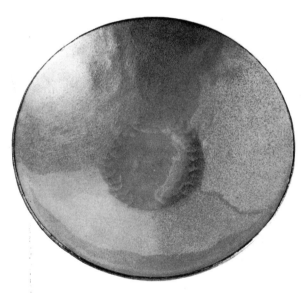

*8* The result is an article enamelled in any colour you choose. It can give pleasure for many years to come. In the following pages different techniques are explained by which an infinite variety of patterns can be created

# I Basic Metalwork

## 1 Cutting and Shaping of Copper

### Grade of Metal

The copper used for enamelling should be a good quality commercial grade. Among the alloys of copper, gilding metal (tombac) is the most suitable for enamelling, but it will stand only a limited number of firings. Copper is supplied in varying degrees of hardness. Use *soft* or *half-hard* material.

### Thickness

For smaller articles, such as jewellery and buttons, 20 or 22 SWG (Standard Wire Gauge) is sufficiently thick. For ashtrays, bowls and similar articles up to 10 in. diameter, 18 SWG is recommended. For articles larger than 10 in. diameter, and also smaller concave articles if they are enamelled inside only, 16 SWG should be used. Copper thicker than 16 SWG is difficult to shape.

Remember that the article will be as strong as the base metal. It is false economy to use metal which does not provide adequate backing. Robust metal is essential to achieve good quality enamelwork of the type discussed in this book. Err on the side of using a heavier rather than a lighter gauge.

### Design

Copper can be worked into almost any shape but there are certain design considerations which have to be observed if the metal is to be enamelled successfully.

*Practical factors:* Balanced and symmetrical shapes are better for enamelling than irregular ones. Circular designs are ideal. Avoid sharp angles, shapes which are too elongated, and those which terminate in points.

*Aesthetic factors:* If elaborate decoration in enamel is intended, keep the shape of the article simple. An elaborate design in enamel combined with an intricately contrived shape will result in a confusing and even vulgar effect. Let either the shape of the object or the enamel decoration predominate; competition between the two will not be a success.

### How to shape copper

This book is about enamelling—the decoration of metal with enamel. Elaborate metalwork is therefore outside its scope. The instructions given are, however, sufficient to produce good copperwork. At the time of writing, ready-made copper shapes are starting to become available from manufacturers and this will make copperworking unessential for the enameller. But many will still want to make their own copper shapes for enamelling. For those who wish to go more deeply into the craft of metalworking a number of excellent instructional books are on the market (see bibliography). Most educational authorities provide evening classes in metalworking or silversmithing, so the metalwork can be done at evening classes and the enamelling at home.

### Tools required

The tools listed below are essential for the simplest forms of metalwork:

Shears (or tinman's snips) to cut the metal
File to smooth the edges, and file card to
   clear the file of copper particles
Boxwood bossing and rawhide mallets
Small steel block and steel plate to provide
   a hard, flat surface

Gas torch
Roundhead steel stakes (with socket if you
   have no vice)
Leather sandbag
Scriber

Tools which are useful but not essential:

Rotary cutter (handy for curved surfaces)
Leg vice or bench vice
Wooden block with hollow gouged out on
   end-grain side

## Making a bowl or plate

To make a circular bowl or plate out of a
sheet of copper about 150 mm (6 in.) square:

Heat the piece to a uniform dull red colour
with a gas torch and quench it in cold water.
This process is known as annealing and makes
the metal ductile and easier to work.

Flatten the copper blank before you start
work on it (9). This is done on the steel block
using a rawhide mallet first on one side of the
blank and then on the other.

Locate the centre of the square of copper by
joining the opposite corners with pencil lines.
Mark the point of intersection of the lines
on both sides with a scriber. Draw the shape
with a pair of compasses and roughly trim by
cutting off the corners and most of the metal
outside the circle, using either shears (with one
arm gripped in a vice if you have one) or a rotary
cutter. Then cut round the line, following it
as closely as possible (10 and 11). Next, smooth
off the edge and correct irregularities with a
file (12). If you hold the copper in a vice for
filing, insert soft jaws or a wad of newspaper
to protect the metal. Finish off with steel wool
to get rid of the dangerous snags at the edge of
the metal.

Hollowing can be carried out over either the
hollowed-out part of a wooden block or a sandbag (13 and 14). It is helpful to have some
guidance when hammering. So, using the
centre point of the piece, draw with compasses
on both sides of the copper shape a number of
circles spaced about half an inch apart. Then
draw a straight line from the centre point to
the edge on both sides. This is the line from
which you start hammering. The mark is particularly important when you do the final
hammering on the stake as it is easy to lose
your starting point.

Start hammering with the bossing mallet on
the periphery and move the piece round until
you have completed the circle. The lines you
have drawn will serve as a guide to the area
wherein the hammer blows should fall.

Place the piece face downwards on a steel
plate to show up the distortion at the edges.
Where the edge does not touch the plate,
remedy this with a few taps with a rawhide
mallet (15). Start hammering within the next
circle and again move the piece round under
regular blows. From time to time correct distortion on the steel plate as described above.

During the process of hollowing you will
find that the copper hardens under the hammer
blows. Heating with a gas torch to a dull red
and quenching in cold water will make the
metal easily workable again. You may find it
necessary to go through this annealing process
several times in shaping a deeper piece.

When the required depth has been obtained,
transfer the piece to a roundhead stake and
move it round as you hammer with the
rawhide mallet within your guide lines (16).
Manipulate the piece so that it fits snugly over
the stake. Give the edge a final hammering
(17). Aim to achieve accuracy with a circular
shape but not absolute smoothness as the slight
irregularities look well under transparent
enamels.

To prepare a small, flat base for a deeper
bowl, mark out on both sides the size of the
circular base you want. Place the article face
downwards and hammer within the circle to
recess the metal slightly with the broad end of
the bossing mallet (18). Turn the article over
and flatten the hump within the circle with the
narrow end of the bossing mallet (19). Finally,
with a few well-placed taps, ensure that the
workpiece sits firmly on the rim of its base.

If you are making a plate with a larger base, it is best to draw in where you intend the base to be and draw your hammering circles outside that line. Do not hammer inside the circle enclosing the base when you do the hollowing or working on the stake.

To prepare the base of a plate, first mark the intended size on both sides. Then place the workpiece upside down on the steel plate and depress slightly with a rawhide mallet the area of the base (*20*). Finally, turn the workpiece over and, with the mallet, flatten the hump formed within the base area (*21*).

*10* Cutting the trimmed shape with shears. One arm of the shears is gripped in the vice to make cutting easier

*9* Flattening the copper

*11* If a large number of curved shapes are to be made, a rotary cutter is a good investment

14

Smoothing the edges with a file

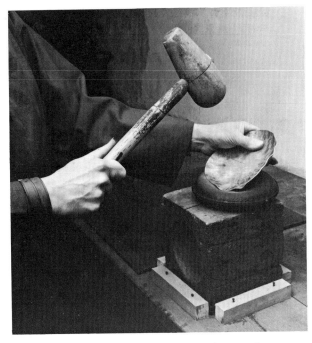

14 Hollowing on a sandbag. Note the wearing of a wrist-strap in metalwork to avoid straining the wrist

3 Hollowing on a wooden block

15 Correcting irregularities on a steel plate

*16* Hammering on a steel stake

*17* Giving the edge a final hammering

16

Recessing the bottom of the bowl for a flat base

20 Making the base of a plate

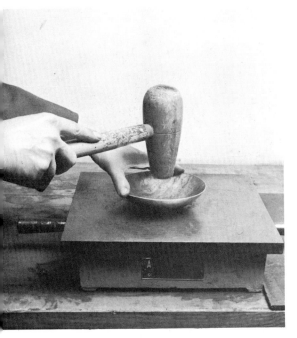

9 Flattening the base from the inside

21 As you have to form a larger base than for a bowl use a rawhide mallet

17

# 2 Etching

Champlevé was the first method of enamelling practised by ancient craftsmen. Originally the metal shape was cast in the metal, leaving a depressed area to be filled in with enamel. Later the depressions were gouged out from the metal with a tool. Nowadays acid is generally used to eat out lower levels which are wet-packed with enamel using the inlay method. The object is to achieve a smooth, even surface, enamel and metal being at the same level. Some modern enamellers allow only a very shallow etch, using the design as a textured effect rather than to separate colours, and covering the whole piece with transparent enamel. The area of metal which is not to be etched is covered with stopping-out varnish, counter-enamel, or candle grease.

Prepare the face of the counter-enamelled piece as for enamelling. A pattern can be traced on the article with carbon paper, or the varnish painted directly on with a brush (*22* and *23*). Charge your brush generously to make it flow freely as you draw, but wipe it from time to time as the varnish dries quickly. If you find it becoming too thick, dilute with lighter fuel or benzine. Do not forget to cover the edges of the piece with varnish. After varnishing all the parts you wish to be protected from acid, leave the workpiece to dry. Cover the exposed base with melted candle grease—it is cheaper than varnish. Then place the article in an acid solution which should be prepared and kept in a glass or acid-resistant plastic container. A strong solution will work faster but the margins will come out rough and the acid may seep under the varnish and etch those parts you wish to be protected. A solution consisting of one part of nitric acid to five to seven parts of water is about right. *Always add the acid to the water.* If you do it the wrong way round, the acid will spit out and burn you or your clothes. Should you get acid on your skin, rinse the affected part under a running tap, wash thoroughly with soap and water, rinse again and apply bicarbonate of soda.

It is not easy to judge the time needed for a piece to be etched to the right degree; the freshness of the solution is an important factor. The best way is to lift the article out of the acid bath with copper or brass tongs, rinse it under running water and examine the effect the acid has had. Check that the varnish is intact and touch up if necessary, again allowing it to dry before replacing in the acid solution. During the etching process some gas bubbles will rise to the surface of the bath and others will remain on the copper. Wipe the metal over with cotton wool held with tongs to disperse the bubbles trapped at the margins between copper and varnish (*24*).

When the etching is completed, rinse the workpiece thoroughly under running water and clean with detergent and water, then dry. The varnish is then removed with cotton wool soaked in petrol (*25*). Continue rubbing until the cotton wool shows no trace of the dark-coloured varnish. Warm with a gas torch to melt the candle grease and pour this into a container for re-use. The remaining candle grease can be removed with hot water. Finally, heat up the workpiece with a gas torch to burn off traces of etching ground and grease. If it has been counter-enamelled, instead of using a gas torch give it a short firing in the kiln.

Prepare the article for enamelling (see Cleaning page 30) but do not use a polishing mop or you may impair the sharp definition of the etched design. Steel wool will not harm it, but it is best to run over the surface with a magnet afterwards to get rid of any particles of steel remaining in the depressions.

Deep etching is suitable for inlay work in the recessed parts and shallow etching for coverage of the whole surface with transparent enamels (*26* and *27*).

(*See figure 163 for illustration of champlevé.*)

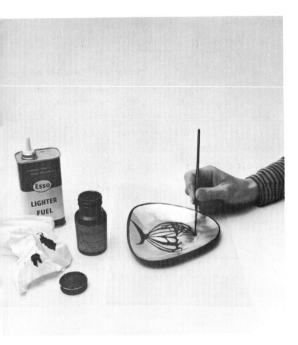

22 Painting the traced pattern before etching

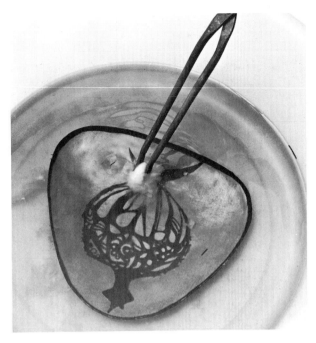

24 Dispersing the bubbles in the acid bath with cotton wool

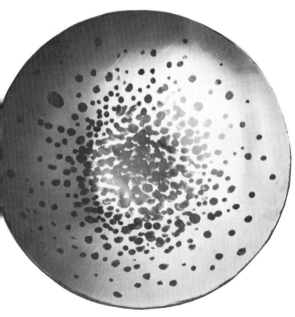

23 Varnish painted freely will give a texture effect

25 The removal of the varnish after etching is a messy job. Petrol will take varnish off your hands

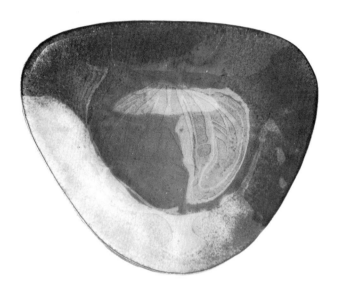

27 A textured effect

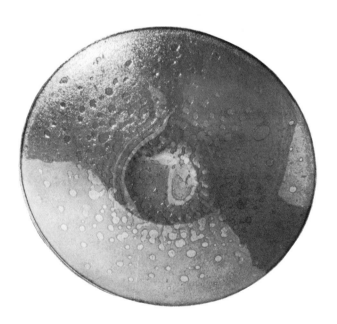

# II Preparations for Enamelling

## 1 Enamels

### Grinding

Jewellery enamels, which are used in enamelling copper, are supplied in 'cake lump' (chunk) or ground form.

If in lump form, they must be ground in a porcelain mortar with a pestle before they can be used for dusting or inlaying. This is a hard, slow job.

The mortar (175 mm [7 in.] dia. is a good size) should be of hard, unglazed porcelain. It needs to stand on a soft pad; a leather sandbag used for copper beating is ideal, but a really thick wad of newspaper is suitably resilient. You will require a wooden or rawhide mallet which is also used in copperwork.

Place about 56 g (2 oz) of enamel in the mortar and pour in water to cover the largest chunks of enamel (28). Having seated the mortar on the leather bag or wad of newspaper, break up the largest enamel chunks by striking the pestle, which is held over them, with repeated sharp blows of the mallet (29). (Do not use a steel-headed hammer otherwise you may break the pestle.) The water prevents the enamel fragments flying out of the mortar. Work the lumps away from the sides and to the bottom of the mortar in order to avoid cracking it by hammering on the sides.

When the chunks are broken up, start grinding with the pestle with a slow, rotary action, pressing the pestle hard against the mortar (30). Some find it comfortable to grind with the mortar held in the lap. If you feel the pestle passing over chunks instead of grinding with a characteristic sound, then there are still lumps to be broken down with the pestle and mallet. After grinding for a minute or so, pour off the now milky and slightly coloured water, taking care not to pour off any enamel from the bottom

28 Chunks of enamel covered with water in preparation for grinding

29 Breaking up the larger chunks before grinding

21

*30* A slow, rotatory action is used in grinding

*31* Large lumps are cracked by heating and plunging them into iced water

of the mortar. Continue grinding, then add a little water, stir briskly with the pestle, and, before the enamel has a chance to settle again, pour off the water into a bowl. Continue with this cycle. The secret of good grinding lies in the frequent removal of the too finely powdered enamel with the milky water and the enamel which is already ground sufficiently in suspension in the water. A good tip is to tilt the mortar suddenly when pouring off the water. This leaves behind the larger pieces of enamel for further grinding. It will be necessary to replenish with water about six to nine times in order to grind 56 g (2 oz) of enamel to the correct size. It should have the appearance of fine sand. Finish by rinsing the pestle and flushing the ground enamel from the mortar.

If the enamel supplied to you is in lumps larger than the size of hazel nuts, preheat your kiln to about 400°C (752°F), place the lumps on a firing rack and leave them in the heated kiln for about 10 minutes. Then take them out of the kiln and plunge them into cold, or, better still, iced water (*31*). They will either shatter to a size suitable to be dealt with in the mortar, as described above, or crack so that they can be broken up easily.

## Washing

It is necessary to wash all enamels, whether you buy them ground or grind them yourself, although in the latter case the enamels will require fewer changes of water.

For washing up to 1 kg (2 lb) of enamel a 1·70 litre (3-pint) mixing bowl or basin serves well. The number of washes needed varies with the type of enamel. Between three and five changes of water suffice for opaque enamels, but transparent and translucent enamels should have from five to eight washes.

Place the powdered enamel in the mixing bowl, run tap water into it and stir vigorously with a one-piece plastic spatula (*32*). After letting it stand for a minute or two, pour off as much water as possible (*33*). You will find the water is milky and slightly coloured. This is because it contains enamel which is too fine for use and also particles of porcelain from the mortar or from the ball mill in which manufacturers have ground the enamel. Tilt the bowl from side to side so as to get rid of as much water, and impurities with it, as you can. The last two washes should be with distilled water. If you are washing more than one colour at the same time, be sure to brush the spatula clean before you start on a new colour so as to avoid colour contamination.

## Drying

The drying of the enamel after washing can be done on top of a heated kiln or in the oven of a cooker. Do not dry in a mixing bowl or basin as the enamel will cake on. The best way is to make a tray on a firing rack with a sheet of aluminium foil. Turn up the four edges of the foil to stop the saturated enamel from spilling over. Spoon the enamel into the aluminium foil tray with a spatula (*34*). You will find that you cannot remove quite all, so add a little water to the residue in the bowl, pour it into a separate container, and keep this with other enamels of various colours for counter-enamelling.

Wash the bowls and spatula carefully to avoid colour contamination before putting them away.

If you have a kiln going, place the firing rack with the aluminium foil tray containing the wet enamel on top of the kiln. Alternatively, the drying can be done in a cooker oven set at the lowest possible heat. Leave the oven door ajar so that steam can escape. A 1 kg (2 lb) batch of enamel takes from 8 to 10 hours to dry but several batches can, of course, be dried simultaneously (*35*). Make sure the powder is quite dry—it should feel like dry sand—before it is graded and stored. Brush off the dry enamel from the foil and discard the foil.

32 Washing the ground enamel. It must be stirred well with a spatula

33 Pouring off the milky and slightly coloured water

34 The washed enamel is spooned into a tray made from aluminium foil

35 Different colours can be dried simultaneously on a large rack

## Grading

After washing and drying, the enamels must be graded into different categories of particle sizes. In the authors' opinion this applies to most enamels bought ready-ground as well as to those you grind yourself. The object in grading is to make quite certain you have the correct grain size for the job in hand. Generally suppliers' standard grinding is 30 mesh and 50 mesh, and 50 mesh is the one to order for both opaque and transparent enamels.

An 80 mesh and a 200 mesh sieve are required. When grading, use some form of filter mask to prevent the inhalation of fine enamel powder. Mechanical filter respirators can be bought quite cheaply or a smog mask will do. If you have a flexible extension to your vacuum cleaner, it is a good idea to suspend this over the dish or tray over which you are sifting to carry off the enamel dust from the atmosphere (36).

Before and after grading, clean the sieves with a stiff brush to avoid contaminating one colour with another (37). Sift over a clean enamel photographer's dish or some other receptacle. Spoon the enamel on to the 80 mesh sieve, gently breaking up with a spoon the lumps of powder which have formed in drying. Do not overload the sieve; small amounts sieve better. A 205 mm (8 in.) sieve takes 5 to 6 spoonfuls comfortably. Too vigorous tapping of the sieve will cause the enamel powder to fly up. Sift thoroughly by tapping the side of the sieve. Some enamel will remain in the sieve. Put this aside. Some of the enamel which has passed through 80 mesh is too fine for dusting and would cause cloudiness in firing. Therefore it should be sifted once again, this time using a 200 mesh sieve. Give the sieve just a few taps and this will get rid of the excess very fine enamel. The powder which goes through 200 mesh can be used as described on page 69. The enamel between 80 mesh and 200 mesh, that is the enamel remaining in the 200 mesh sieve, is suitable for dusting and inlaying.

*36* Grading the dried enamel. The flexible extension of a vacuum cleaner is switched on, and a respirator is worn to prevent the inhalation of the fine enamel dust.

*37* The sieve is cleaned between each colour by being turned upside down and cleaned with a stiff brush

Returning to the coarse enamel left in the 80 mesh sieve, this can be used for special effects (see page 67). It should be sifted through a coarse metal sieve to eliminate the small lumps of very fine powder that form in the drying; these lumps should be thrown away. See drawing A.

With a thorough washing you can expect to lose about 14 g ($\frac{1}{2}$ oz) in a 1 kg (2 lb) batch, but on the other hand your care will be rewarded by clear, sparkling colours in your enamelling.

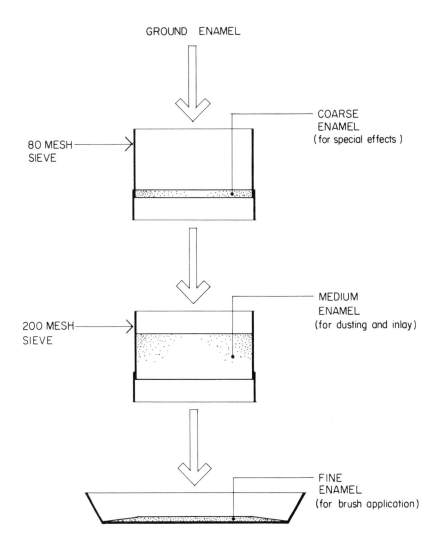

GROUND ENAMEL

COARSE ENAMEL (for special effects )

80 MESH SIEVE

MEDIUM ENAMEL (for dusting and inlay)

200 MESH SIEVE

FINE ENAMEL (for brush application)

A

## Storing

You will now have three grades of well-dried ground enamel, all of the same colour—coarse for special effects, medium for dusting and inlaying, and fine for brush application. Store them in clean, dry, screw-top glass jars and label them with the manufacturer's name and reference code, colour, grade, and firing temperature (*38*).

There may be steel particles in any grade of enamel. These can be extracted by placing a small magnet in a funnel and passing the ground enamels through a few times (*39* and *40*). Steel particles cause dark brown specks in your enamelling.

Moisture from the air will cause ground enamels to deteriorate, so keep the jars, with lids firmly screwed down, in a dry cupboard or box.

*38* Any clean, dry, screw-top glass jars can be used for storing. Labelling is essential for identification and re-ordering

Passing the ground enamel over a magnet to get rid of steel particles

*40* The magnet's haul. Steel particles in enamel cause dark brown specks after firing

# 2 Gum

Proprietary gum solutions can be bought, but it is a comparatively simple matter and much cheaper to prepare your own solution using gum tragacanth powder.

The making of gum solution can be done in the kitchen as gum tragacanth is harmless; it is, in fact, an ingredient of some food products.

The powdered gum tragacanth is mixed with distilled water in a bowl or saucepan—any grease-free utensil will do. Use one heaped tablespoon of gum to a quart of water. First sieve the powder through a metal kitchen sieve to break down the lumps (*41*) and then mix to a paste with a little water (it dissolves more easily this way) (*42*), before adding the remainder of the distilled water. Stir with a spoon while adding the water to break up the lumps. Bring slowly to the boil. Holding the sieve in the solution, work away as many lumps as possible with a spoon (*43*).

The solution has then to be strained and bottled. A plastic funnel lined with a double thickness of old nylon stocking makes a good strainer. Place the metal sieve, which has been washed free of lumps, in the lined funnel.

Spoon in the solution but do not force it through the sieve (*44*).

You now have a concentrated solution of gum which requires diluting to suit the different methods of application. For *spraying and inlaying*, use about two dessertspoons of concentrated solution to a pint of distilled water. A correctly diluted solution is slightly cloudy but can only be seen to be so when held against a bottle of clean water. The solution is a little tacky to the touch, but not more so than saliva. *For brushing*, a slightly more concentrated solution is required so use three dessertspoons of the original concentrated solution to a pint of distilled water.

Always shake the concentrated solution before measuring for dilution and also shake the bottle of diluted solution before use, as the thick portion tends to descend to the lower part of the bottle.

It will be found that one mix of gum solution will go a long way. It keeps best in a refrigerator or a cool place. It is worthwhile making a fresh mix after about two months.

It is important that the solution for spraying and inlaying should not be too concentrated otherwise it will cause cloudiness in colours, particularly transparent ones.

The dry gum tragacanth is broken up before being mixed with water

*43* Lumps have to be worked through a sieve while heating the solution

Add water little by little, stirring all the time

*44* Strain well before bottling

# 3 The Copper

## Cleaning

Before it is enamelled, copper requires three separate treatments for cleaning: heating, pickling, and polishing. Partially enamelled pieces require pickling and polishing only.

## How to clean the copper

*Heating* can be done either in a kiln or with a blow torch over a sheet of asbestos (*45*). The metal workpiece is heated up to a dull red glow and, with tongs or pliers, plunged straight into a bucket of water. Before quenching the piece in water, make quite sure all parts of it are evenly heated.

*Pickling* is done in a solution of nitric acid and water. *Always add the acid to the water*. If you do it the wrong way round, a violent chemical reaction takes place and the acid will spit out and may burn you or your clothes. Mix one part of nitric acid to five parts of water. The container may be either glass or acid-resistant plastic. Keep a lid over the acid bath and do not lean over the container or breathe in the fumes. The acid solution should be kept and used in a well-ventilated room and must be out of the reach of young children.

Place the cold copper article in the acid solution using brass or copper tongs (*46*). Rinse the tongs under running water after they have been in the acid solution. Do not allow the acid to get on your fingers, but if it does, rinse them under a running tap, wash with soap, rinse again, and apply bicarbonate of soda.

Leave the copper in the acid until it has turned a pink colour. The time the acid solution takes to work depends on the freshness of the solution. A new solution will treat the metal in a matter of minutes if it has not been fired. After firing, however, when there is fire-scale present, it will take considerably longer. Remove the piece with tongs and rinse well under a running tap. Scrub with a stiff brush— a nylon washing-up brush will do—using a liquid detergent to remove any traces of grease and acid (*47*). Do not touch the surface you intend to enamel or you will leave fingermarks on the metal. You will soon get used to holding articles by the edges or on a side which is not immediately to be enamelled or has been enamelled already. The whole cleaning process is aimed at producing a grease-free surface, as neither gum nor enamel will adhere where there is grease. Dry with a clean cloth.

With partially enamelled pieces you may find that some colours are affected by the acid causing discoloration or a matt, powdery surface. Do not worry about this, the next firing will restore colour and gloss.

The pickle solution is colourless at first. After use it turns bluish but this does not indicate there is anything wrong with it. When you find pickling is taking much longer, the solution can be strengthened by adding more acid, but when it weakens again, discard it. When it has reached this stage you can add water to the solution. Dilute it with as much water as your container allows and then pour it down the drain and let the tap run for a while.

As an alternative to cleaning with acid a solution of ordinary vinegar and salt can be used. Just add a couple of dessert spoons of salt to about 125 ml ($\frac{1}{2}$ pint) of vinegar and stir well until there is a frothiness on the surface. The solution tends to work best after it has been used a couple of times; on average it takes about ten minutes to clean off firescale and a soaking of several minutes is adequate when first preparing the metal for enamelling.

The metal must be *polished* to make it bright and shining for enamelling. This can be done by using steel wool (fine in order to avoid scratching the surface deeply) (*48* and *49*). Gloves are recommended for this work as the small pieces of metal from the steel wool can enter the skin of the hands with painful effects. Gardening gloves are suitable.

Rub the copper with a circular motion until it shines; the fine scratches give a good reflecting surface which looks beautiful when enamelled.

After polishing, remove all specks of steel wool by brushing with a clean brush and passing a magnet over the surface.

If you have a polishing motor, or an electric tool which can be fitted with a polishing mop, this will save you work (*50*). Greaseless polishing compositions are obtainable for use with polishing mops.

Finally, it is advisable to play the flame of a blow torch over the surface immediately before enamelling to make sure it is quite free of grease and other impurities (*51*).

*45* Heating the metal to a dull red glow before plunging into cold water.

*46* Removing the firescale by pickling

*47* When scrubbing, take care not to touch the surface which is going to receive the enamel

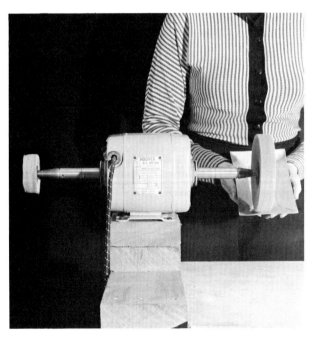

*50* A polishing motor saves time and energy

*51* Playing the flame briefly over the workpiece

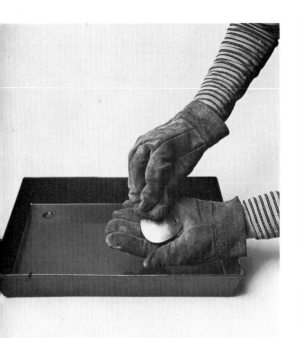

Small pieces can be polished in the palm. Gardening gloves prevent particles of steel wool entering the skin and also protect the polished surface from finger-marking

*49* Larger pieces are easier to handle. A bigger wad of steel wool is needed

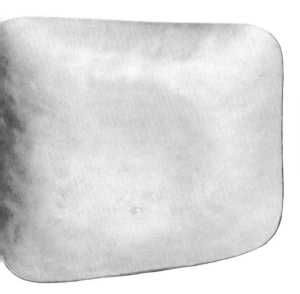

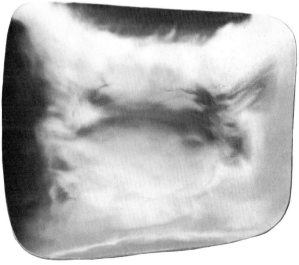

Copper is a dull pink colour when it comes out of the pickle

*53* After polishing it shines

33

# III Application of Enamels

## 1 General

There are three main types of enamel: opaque, opalescent or translucent, and transparent. Flux is a clear, colourless, transparent enamel.

Manufacturers usually indicate firing temperatures either by stating that these are around a certain temperature, or by indicating a range with maximum and minimum temperatures. Within the same range of firing temperatures there are soft, medium, and hard enamels. Soft enamels fuse at the lowest temperature, hard enamels at the highest, and medium enamels in between the highest and lowest. All three grades of enamels can be used on one piece at the same firing.

If manufacturers do not give firing temperatures for their enamels, enquire before using them in order to avoid the disappointment of a bad result.

### Thickness of Coating
#### Transparent enamels

Whatever technique you are using, the thickness of the unfired first coat should be a little greater than is needed to cover the metal. If, however, the coat is too thick, the enamel when fired will become cloudy, or small craters, giving an orange-peel effect, will appear on the surface. Both these defects will show up after later firings. If the coat is too thin, then small holes will be burnt into the enamel, exposing the metal surface covered with thick firescale. As some enamel is present in this firescale, it cannot be removed entirely by pickling.

#### Opaque enamels

Opaque enamels need a slightly heavier coat than transparent enamels; but, again, if the coat is put on too thickly, you will get an orange-peel effect and too thin a coat will cause burnt-out holes.

With both opaque and transparent enamels, if you want a rich colour, resist the temptation to put on a thick coat. Instead apply two normal coats in two firings.

With both transparent and opaque enamels, the thickness of subsequent coats can vary according to the effects you want to achieve. They can be thinner than the first coat but avoid too heavy application. There should always be a carefully applied first coat of the correct thickness.

### Evenness of coating

It is most important to obtain an even first coating of enamel. If you do not, patches of enamel may burn out where it is too thin and flake off where it is too thick. Colours, especially with transparent enamel, show up unevenness and the thick parts can be orange-peeley and cloudy.

### Counter-enamel

Generally speaking, it is necessary for an article to be enamelled on both sides. This is because copper, when heated, expands and draws the enamel with it. When it cools, the enamel solidifies before the metal has fully contracted with the result that the metal shrinks away from the enamel. These stresses can distort and warp the metal. In some cases, especially with flat or convex shapes, the stresses which have developed in the metal can crack the enamel or cause it to fly off. Enamelling an article on both sides helps to balance these stresses. Applying a coat of enamel to the

34

back of an article is known as counter-enamelling.

Convex shapes can be enamelled only on the inside, provided the copper is of a heavy gauge and the shape is circular or nearly so. Irregular shapes, elongated shapes, or shapes with sharp angles should always be counter-enamelled.

Always apply the counter-enamel first. Ideally the thickness of the coat of counter-enamel should be the same as that of the overall coating on the face, but in cases where several coats are applied this will, of course, be impracticable. There is, however, no need to be so careful about keeping down the thickness of the counter-enamel, and it can be put on in one thicker coat using a suitable mixture of enamels. Repeated firings will eliminate defects which may arise.

A mixture of colours produces a good tough coat and makes a more satisfactory counter-enamel than any single colour. A composition which has proved satisfactory in practice is 50 per cent hard flux and 50 per cent mixed enamel colours, mostly transparent. Avoid opaque reds and oranges. Counter-enamelling is a good way of using up left-over enamels.

If you are enamelling a deep vessel, such as a vase or bowl, and wish to develop a design on both the inside and outside using unmixed colours, do not try to finish one side at a time; work on the two sides alternately.

Do not attempt to enamel the metal edges as when enamelled they are liable to be damaged. A polished copper edge looks very well on enamelled pieces.

As you must never touch a surface which has been prepared to receive a coat of enamel, it is advisable to make a 'dry run' before cleaning to find out the best way of holding the workpiece. Pieces which have the counter-enamel already fired on can be held in the palm of the hand on the counter-enamelled side.

# 2 Methods of Covering the Copper Surface with Enamel

There are two basic methods of covering the copper surface with enamel. One, known as *dusting*, is by sifting on the enamel as a dry powder and the other, *inlaying*, by applying it in the form of a wet paste. A gum solution is needed for both methods to make the enamel adhere to the copper in the dense, integrated coat which is essential for good results in fusing.

## Dusting

When the enamel is dusted on, there are two methods of applying the gum solution. The simple one, although not the most efficient, is to *brush on* the solution. A soft 15–20 mm ($\frac{1}{2}$–$\frac{3}{4}$ in.) flat camel or squirrel hair brush is all you need.

The other method is to *spray on* the gum solution, and for this there is a choice of equipment:

electric sprayer (*54*)
spray gun operated by compressor (*55*) (a second-hand one can often be obtained for a few pounds)
mouth sprayer (ask in artists' materials shops for 'fixative atomiser')

All of these types of sprayer do the job satisfactorily. A mouth sprayer is suitable for small objects in small quantities only, as it is tiring and can be troublesome in use. For the serious amateur and professional a good spray gun and compressor with air tank, or an electric sprayer, is essential.

For the dusting process itself, a sifter is required. A plastic tea strainer with fine mesh (80 or a little larger) is quite adequate for the purpose. Proprietary sifters specially made for enamelling are available from suppliers.

The technique of *brushing on* the gum solution is only suitable for up to 75 mm (3 in.) and more or less flat articles (*56*). When brushing

54 A small electric sprayer is very handy for spraying gum solution

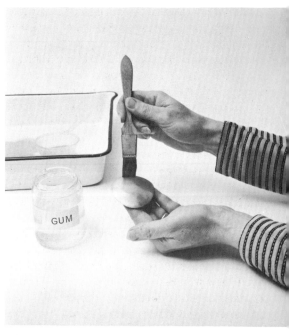

56 Painting on the gum solution is suitable for small pieces only

55 Spray gun operated by compressor

57 Touching up bare patches at the edge

on the gum, use a fairly concentrated solution (see page 28) and cover the copper surface thoroughly. The gum dries quickly so it is important to work rapidly with both brush and sifter. Apply a generous coat of solution, leaving the highest part of the article until last or the solution will run down to the lower part. Before the gum has a chance to dry, dust on a complete covering of enamel. When you have finished dusting, the enamel surface should be damp. Bare or dry patches are repairable at the edges only. To touch up with more gum solution, press the brush charged with gum against the edge of the workpiece and tilt it so the portion to be patched is uppermost (57). Sift on more enamel if necessary. A bare patch in any place other than at an edge is difficult to rectify and it is better to wipe off the enamel, saving it for counter-enamelling, and start again. One difficulty with brushing is the tendency of the gum solution to run when the piece being enamelled is deeply concave. It helps to tilt the article as the brushing is done and, again, to work quickly. After use, rinse the brush in water.

When *using a spray gun*, adjust the nozzle to a fine spray. The first application of gum should be light so as to give a dewy effect (58). Dust on sufficient enamel to cover the gum (59) and then re-spray with gum over the whole surface and follow with a second layer of enamel. The enamel coating must be dense and compact, otherwise trouble will be experienced when it is fused. Compactness is achieved by keeping the enamel thoroughly moist. Continue spraying and dusting alternately until sufficient thickness of coating is achieved (60). Finish off with a light spray of gum solution.

When using a compressor with a spray gun, a pressure of 20 to 22 kg/sq. cm (45 to 50 lb/sq. in.) will produce a satisfactory spray.

Do not spray over your working surface. Turn sideways from your table and spray downwards. Protect the floor with layers of newspaper. The gum solution is not harmful but makes the floor sticky and slippery.

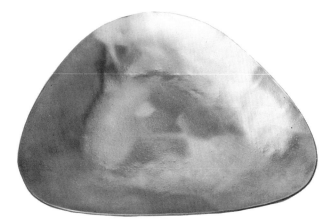

58 The light first spray

59 Just enough enamel dusted on to absorb the moisture

60 The full covering coat

37

## Points to be watched in dusting

Keep the bottom of the sifter covered with enamel—about one-quarter full is best. If the bottom is not covered, you cannot sift evenly; and if the sifter contains too much enamel, its weight will prevent the powder coming through the mesh.

Dust over a tray so as not to waste enamel; an enamel photographer's dish is excellent but a clean sheet of paper will do. Enamel collected on the tray or paper can be returned to its storage jar.

Each time you lift the sifter, tap it once over the tray before you start dusting on the workpiece. Enamel may be clinging to the bottom of the sifter, and this can cause blobs which will show up after firing. This preliminary tap before dusting will become a habit and save you a lot of annoyance later.

Release the enamel on the workpiece by tapping the sifter gently.

Hold the surface being treated at right angles to the falling enamel, tilting the workpiece as necessary (*61* and *62*).

(*63* and *64*)

*61* The workpiece is turned and held so that
*62* the enamel always falls on the surface at right angles

38

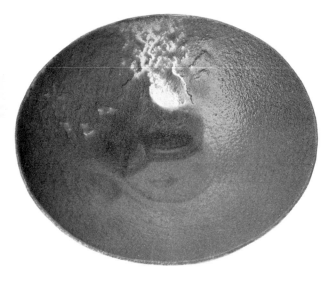

*63 and 64* Alternate spraying and dusting are important.

The first picture shows how too much gum washes away the enamel. The second picture shows too much enamel dusted on; the next spraying will blow off the enamel and cause uneven coverage. In both cases the enamel should be wiped off and the piece rinsed and prepared again to receive a coat of enamel. As it has already been polished, only a very light polishing will be necessary

## Inlaying

Three simple tools are needed for inlay work: spatula scriber spreader.

These can be purchased from suppliers, but they can also be made quite easily as they consist of nothing more than shaped copper wire inserted in a wooden handle a little thicker than a pencil. See drawing B.

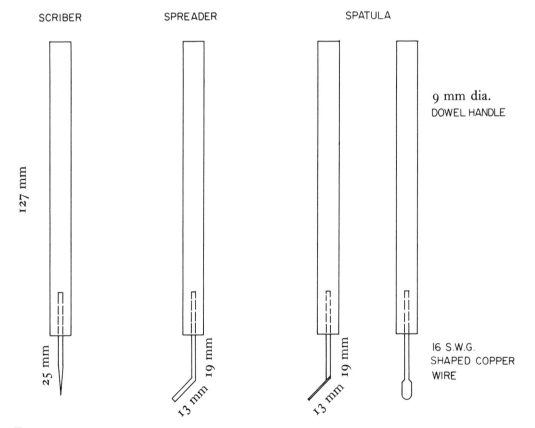

SCRIBER  SPREADER  SPATULA

9 mm dia.
DOWEL HANDLE

127 mm

25 mm

13 mm   19 mm

13 mm   19 mm

16 S.W.G.
SHAPED COPPER
WIRE

**B**

With this traditional method of enamelling metal, the powdered enamel is soaked with gum solution and applied to the copper surface as a paste.

Place some enamel, of the same mesh as for dusting, in a small dish (an artist's colour dish is ideal) and soak it well with gum solution (for suitable concentration see page 28). Prop up the dish so that the excess solution drains away from the enamel. Pick up the paste with a spatula (65) and transfer it to the workpiece with a scriber (66). Use a spreader to even out the coating and pack it tightly (67). This method of laying on an enamel coating is ideal for smaller multi-colour pieces, jewellery, and buttons. The reason for not employing this method on large pieces is that it is rather laborious and slow.

66 The scriber is used to transfer it to the workpiece

5 In inlay work a spatula is used to pick up the enamel paste

67 The spreader evens it out

# 3 Firing

Firing, which bonds the enamel to the metal, is the climax and the essence of enamelling, for it is this operation which transforms the separate particles of enamel into a smooth, cohesive coating. The fired coating is quite different in appearance from the enamel before it is fired, and applying the heat which brings about this almost miraculous change is without doubt the most interesting phase of the whole process.

Originally the heat for enamelling was provided by wood or coal fires, but the development of electric kilns has now made life much easier for the enameller. These kilns are easily controllable, clean (this is essential for good enamelling), and simple to operate.

For firing smaller articles the flame of a gas torch can be used.

It is worthwhile watching the different stages of the process of fusing through the peephole of the kiln, or as you apply the gas torch; it takes only a few minutes. The sequence of events is roughly as follows:

*First coat:* After the first few seconds the enamel starts to darken in colour and the grains soften and coalesce. The surface becomes rough. As it heats up, the whole workpiece turns red and the surface smooths out. A piece fired to the correct degree has a bright red glow and a smooth, glossy surface. After removal from the kiln, the red glow changes to a dark colour. Within seconds the colours start to return one by one. The constant colour change, which continues until the piece has cooled, is fascinating to watch.

*Subsequent coats:* You will probably hear the previous coat of enamel crack. This is caused by the copper expanding while the enamel is still hard. Do not worry; as the enamel fuses the cracks will heal. The sequence of events in the fusing and colour change is the same as with the first coat.

## Kilns

Special kilns are made for enamelling, but you can enamel successfully in any small, front-loading pottery kiln which has a single door (*68* and *68a*). You must be able to open the door with one hand so that the other is free to wield the firing fork.

The elements are not covered in a pottery kiln. Therefore for safety reasons it is advisable to switch the kiln off before opening the door and switch it on again when you close it. The resulting heat loss is negligible.

The running cost of a kiln is not excessive and some kilns consume as little as a one-bar electric fire. Both enamelling and pottery kilns are available in a variety of sizes. Do not make the mistake of starting off with too small a kiln. (A gas torch or hotplate kiln is adequate for small pieces.) On the other hand, do not go to the other extreme and buy a very big one. It is not only the size of the kiln which limits the maximum size of the article you can enamel; the weight of the article is also a factor. Remember you have to open the door with one hand and lift the piece to be fired with the other. A 30 cm (12 in.) bowl of 16 SWG copper weighs about 1·02 kg ($2\frac{1}{4}$ lb), to which must be added the weight of the enamel, rack, and support. The total weight has to be balanced safely on the end of a long-handled firing fork. The weight consideration does not, of course, apply with industrial kilns which are partly or fully mechanised.

It is strongly recommended that you have a pyrometer with your kiln, whether it is a pottery or an enamelling one, for you must know the temperatures at which you are working, if you are to have satisfactory firing control.

Although the firing process takes only a few minutes, it takes place at high temperatures (suppliers give the fusing temperatures of their enamels), so the kiln must be brought up to the correct temperature beforehand. It is advisable to keep it at this temperature for about half an hour before the first firing so that the bricks ab-

An enamelling kiln (right) and a pottery kiln which can be used for enamelling

68a  Firing in a pottery kiln

sorb and store the heat to compensate for some loss when the door is opened. The heating-up time varies according to the size and type of the kiln; but it runs into hours. Smaller enamelling kilns for craft enamelling heat up fast but require time for heat recovery between firings. Pottery kilns heat up more slowly, but, after the period allowed for heat absorption, you can do continuous firing in them. Kiln manufacturers supply data concerning heating up times for their kilns.

A timer will relieve you of the need to devote your attention to your clock or watch, and will leave you free to observe the firing or attend to other matters.

To allow the heat to circulate around pieces, they should be fired on heat-resistant nichrome wire mesh racks. These racks, which should be chosen to suit the size of the kiln (leave 25 mm clearance all round), enable workpieces to be lifted in and out by means of a fork, the prongs being inserted under the racks (69).

A pair of asbestos mittens is required when working with a kiln.

69  Enamelling fork, racks, stainless steel supports and spurs

70 A small counter-enamelled piece fired in a cradle

72 Using a stencil when counter-enamelling so as to preserve a bare metal base. In subsequent firings no support marks will be made on this base.

71 Larger counter-enamelled piece on a cradle

73 The use of spurs as supports on a bare metal base

44

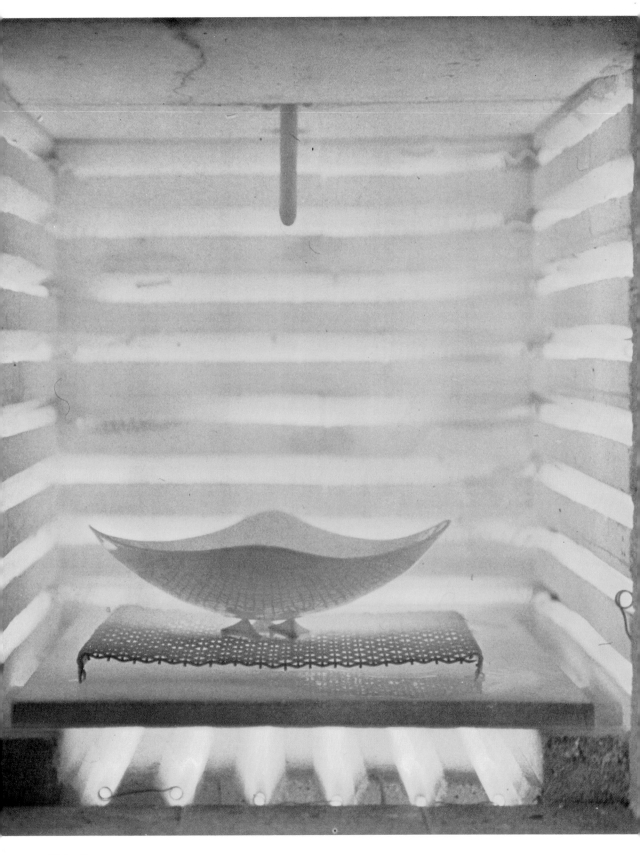

Firing in progress
Photograph taken through spy-hole of kiln

Sheherazade    76 cm x 52 cm (30 in x 20 in)
Marit Guinness Aschan
Kansas Museum of Art, USA

Since enamel becomes soft and sticky when fused, the workpiece should be supported in the kiln, in such a way that contact with the support causes the least possible damage to the counter-enamel. Support marks on the back of an article which has had several firings can spoil it. Decide on the type of support you are going to use before putting on the counter-enamel.

A cradle support, or trivet, is suitable for shallow pieces and for firing counter-enamel (70 and 71). When firing shallow pieces in a cradle, file back the counter-enamel from the edges before preparing the face for enamelling. This type of support can be made of stainless steel secured with stainless steel bolts. As special shapes call for special supports, you may have to improvise.

For deeper pieces a flat circular base of bare metal is recommended. To obtain this base, first decide the diameter of the smallest circle on which the workpiece will stand safely when supported at three points. 76–100 mm (3–4 in.) dia. pieces will need 30 mm (1¼ in.) bases; a 38 mm (1½ in.) base is suitable for up to 150 mm (6 in.) diameter pieces. Bear in mind the fact that flat pieces will need larger bases than deep ones as they tip over easily if supported too near the centre.

Draw the size of the base on the article with compasses, using the centre point. It is important to have the base in the middle not only for appearance but also for balance when moving the piece in and out of the kiln. Cut out a stencil (see page 53) of a size equivalent to the intended size of the base and soak it in gum solution. Place it on the base of the article within the circle (72) and apply the counter-enamel. Remove the stencil in a half-dry condition. By supporting the workpiece on this base in subsequent firings you avoid unsightly marks on the counter-enamel.

In further firings, even if you have an exposed copper base, you should not place the workpiece directly on the rack. Supports known as spurs will support it clear of the rack. Place the three-pointed sides of the spurs downwards so they have a firm foothold. Use three spurs to support the workpiece (73). You will find it is convenient to position and mount it on the spurs at eye level.

Make sure the article is standing firmly on its support, or supports, before firing.

With small pieces you can fire more than one at a time. The number depends on the size of the kiln and rack (74 and 75).

74 Small pieces fired together on a special rack

75 Small pieces fired together supported individually on spurs

After counter-enamelling, and also after the firing of the first full covering coat on the face, the workpiece should be placed upside down and a weight placed on the base. This helps to straighten the edges and correct the effects of warping in cooling.

This is a job which must be carried out quickly and one which requires judgment. The time to do it is just as the red glow leaves the fired piece and before the enamel starts to cool. Weighting after counter-enamelling presents no difficulties as the piece has been fired face downwards. All you have to do is to lift it from the support—a paint scaper makes a good lifting tool—place it on a hard, heat-resistant surface and quickly put the weight on it, pressing it down. After the first coat is fired on the face, lift the article off the supports (76) put it down, and, gripping it with flat-jawed pliers (77), turn it over (78) and put the weight on it, pressing it down (79). You will soon learn to judge the right moment to carry out the operation. If you do it too soon, the paint scraper or the pliers will mark the still-soft enamel. If you do it too late, the hardening enamel will crack. However, if the marks or cracks are slight, they will heal in the next firing.

76 Everything should be handy for the lifting-off operation. The workpiece is taken from the supports as soon as it loses its red glow

77 Grip the workpiece with pliers

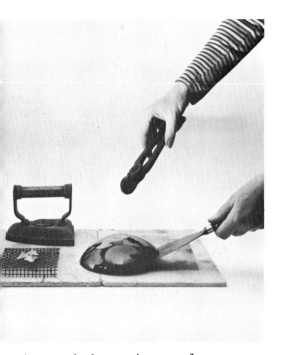

The top of the kiln is an excellent place for drying the enamel coat before firing.

*Hotplate kilns* are small, simplified versions of enamelling kilns (*80*). Their advantages are their low price and the speed with which they reach firing temperature. They are very suitable for jewellery and other small pieces. Normally you fire these pieces unsupported, so remember that in this case they must not be counter-enamelled. As hotplate kilns are not fitted with pyrometers you fire each piece until the surface is seen to be smooth and glossy.

Hotplate kilns are ideal for swirling (see page 73) as you are able to see exactly what you are doing, and the results.

ase it on to the heat-resistant surface

n old-fashioned flat iron is ideal for ressing down the workpiece

*80* A small hotplate kiln suitable for firing small pieces

## Gas torches

An even temperature is necessary in firing, and the size ot the piece which can be fired is therefore limited if a gas torch is used. The article must be small enough for the flame to reach all parts of it at once. Pieces fired by torch cannot be counter-enamelled as they have to be placed directly on the racks. Too much heat would be lost if supports were used.

The article is placed on a rack standing on top of a couple of bricks and the flame is directed from below (*81*). If the flame is allowed to play on the enamel, it will discolour it.

It is a good idea to have a lid to conserve the heat. An old saucepan, which has been scrubbed free of all traces of grease, turned upside down will do. This should be held close above the workpiece. It is most useful to make a small hole in the side of the saucepan to enable you to watch the fusing process without having to lift the saucepan.

The firing is completed when the surface of the piece is smooth and glossy and has a red glow.

## General

Make sure that the coat of enamel is completely dry before firing and that the pieces are handled carefully. The enamel rubs off easily at this stage.

After being fired, the workpiece should be placed on a heat-resistant surface and allowed to cool slowly. Do not leave it in a draughty place.

Beware the prongs of your fork after it has been in the kiln; they will be hot.

Remember that workpieces retain their heat for some time after firing. The return of colour does not mean they are cool enough to handle. Always test for temperature by holding your hand near a piece before picking it up.

On thin gauge copper, such as domed pieces of jewellery, cracks may appear on the counter-enamel at the edges in cooling. These will heal without leaving any blemishes in the next firing, when the other side receives a coat of enamel.

If after firing you find burnt-out holes in the enamel coat, caused by too thin or uneven application, these can be repaired. Prepare the surface for the next coat. Fill the holes with the appropriate colour using the inlaying technique. Give the whole surface a light dusting. If it is a deep object, dust it more liberally on the upper parts as the enamel tends to run away from the edges. Dry the piece and then re-fire.

*81* A gas torch which is also only suitable for firing small pieces

48

## Rough guide to firing times and temperatures

The table given below is intended merely as a rough guide. It is not possible to lay down hard and fast rules for the length of firing and temperatures. Kilns vary and so do enamels and also individual requirements for particular finishes.

The temperatures given apply to the enamels used by the authors and the firing times produce a completely smooth, glossy finish.

The length of firing is affected by two factors causing heat loss:

1 The opening of the kiln door.
2 The amount of heat absorbed by the cold workpiece.

The first factor is the more significant, and therefore all pieces up to 25 cm (10 in.) diameter are fired for the same length of time.

Larger pieces and heavier gauge copper need proportionately longer time.

The firing temperature of the enamels used by the authors is 850°C (1562°F) and this is adhered to strictly. There are, however, two exceptions:

Silver and gold foil burn out easily, so the temperature is lowered and firing time shortened, allowing just long enough for the cracks to heal. (When the foil is covered with enamel in subsequent firings it no longer needs special treatment.)

With overfiring, no matter what the firing temperature of the enamel, the normal firing temperature should be raised by 100°C (212°F).

The temperatures in the table are given in Centigrade, with Fahrenheit in brackets, and the firing times in minutes.

The copper thickness is in British Standard Wire Gauge.

| | Temperature C F | Up to 25 cm (10 in.) diameter 20 SWG and 18 SWG | 25 cm (10 in.) diameter 16 SWG | 30 cm (12 in.) diameter 16 SWG |
|---|---|---|---|---|
| Counter-enamel | 850 (1562) | $3\frac{1}{4}$ | $4\frac{1}{4}$ | $4\frac{3}{4}$ |
| Hard flux or transparents as first coat | 850 (1562) | $3\frac{1}{4}$ | $4\frac{1}{4}$ | $4\frac{3}{4}$ |
| Opaques as first coat | 850 (1562) | $2\frac{1}{2}$ | $3\frac{1}{2}$ | 4 |
| Subsequent firings | 850 (1562) | $2\frac{1}{2}$ | $3\frac{1}{2}$ | 4 |
| Final firing | 850 (1562) | $2\frac{3}{4}$ | $3\frac{3}{4}$ | $4\frac{1}{4}$ |
| Silver and gold foil on pre-fired coat | 800 (1472) | 2 | 3 | $3\frac{1}{2}$ |
| Overfire | 950 (1742) | $2\frac{3}{4}$ | $3\frac{3}{4}$ | $4\frac{1}{4}$ |

# 4 Developing Design in Colour using Basic Techniques

## Dusting two or more colours

Pleasing results can be obtained by dusting on two or more colours. A pepper and salt effect will appear where the colours meet; this will be more noticeable when using opaque or contrasting transparent enamels. Try to achieve an even overall thickness of enamel, taking particular care where different colours overlap. With transparent enamels, apply the light colours first.

Get into the habit of cleaning the tray carefully when you change to a fresh colour so as to avoid contaminating your colours. A vacuum cleaner brush attachment, a large, soft brush, or a non-fluffy rag should be used.

When you develop a design solely by dusting on different colours, the whole surface must be covered by the first coat.

More colours can be added partially where required after the first coat has been fired. While, except with special painting colours, it is not possible to create a new colour by mixing two ground enamels, you can create an endless variety of new colours by firing transparent enamels on pre-fired opaque or other transparent coats. For example, transparent blue over yellow will give a green; blue over red, a violet; and green over red, a brown.

## Sgraffito

Sgraffito is an Italian word meaning *scratched* and consists of scratching off unfired enamel so as to form a design. It is essentially a freehand technique for creating line design in different colours. As it is impossible to transfer a design direct to the object, you either develop your composition as you go along, or follow as accurately as you can a drawing you have made beforehand. It is thus a technique of creating the unique.

Apply a coat of enamel to the copper by the dusting method, but, instead of the final spray of gum solution, finish with a light dusting of enamel, to avoid the formation of a crusty surface which would break away as you scribe your design. Dry the article well and carry out your design by scratching away the powder from the metal with a scriber (*82* and *83*). Be careful not to touch the dusted surface as you work: the enamel rubs off easily. Tip off the loosened powder from time to time and, when you have finished, blow away lightly any which clings to the exposed metal. The piece can then be fired (*84*). After firing, pickle in nitric acid solution to remove firescale from the exposed copper and prepare the article for the second coat in the usual way. For the second coat, apply one or more transparent colours taking full advantage of the different colour effects you will get from each colour over the exposed copper surfaces and over the pre-fired enamel areas (*85*).

Opaque enamels can be used to the best advantage if the sgraffito is done in the second coat over a pre-fired first coat of opaque or transparent enamel.

A ragged-edged appearance can be achieved by scratching your design into the enamel in a half-dry state. This technique is recommended only if you used sgraffito in the second coat. You can also if you wish add smooth lines to the piece after drying it.

On the other hand, as a contrast, very fine, sharp lines can be obtained by dusting 200 mesh enamel through a 200 mesh sieve on a pre-enamelled surface. Do the scratching with a needle point.

(*86*, *87*, *88* and *89*)

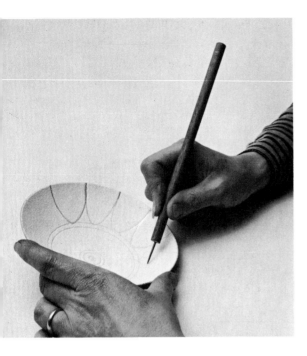

*82* The outline of the design being scratched in the dry enamel. The tool used is a scriber which is also used in inlay work

*84* After firing, the exposed copper is covered with firescale

*83* For larger areas it is worthwhile making a special tool. A piece of copper wire is hammered flat and cut straight at the end

*85* When the piece is dusted, the design shows through the powdered enamel

86 Finished piece in sgraffito. It has had two full covering coats and two firings. The same colour looks quite different over copper and pre-fired enamel

87 A more elaborate design on a bigger piece

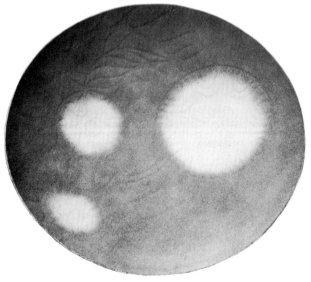

88 Two or more colours can be used for covering

89 Sgraffito can be as simple or as elaborate as you wish

# Stencil

Stencilling is a method of confining enamel to sharply defined areas. It is excellent for repetitive work and is also used in enamelling individual pieces. The part of the article you wish to be free of enamel is covered by the stencil.

The stencil itself should be of absorbent paper so that the gum solution will not run. Paper towelling which retains its strength when wet does well, or you can use blotting paper provided it remains fairly strong when soaked. Stencils can be kept for re-use.

Draw your design on paper (keeping in mind that stencil paper will stretch when wet), then place a sheet of carbon paper in between the drawing and the stencil paper and trace over the drawing with a pencil. Cut around the carbon impression on the stencil material with a stencil knife or scissors; a stencil knife is better as it is easier to operate where you have an intricate design, and, provided that you use a sharp blade, makes a clean, sharp-edged cut. When using a knife it is as well to have a cardboard backing as this will give a clean edge and also prevent the blade cutting through to the bench or table on which you are working (90). A knife is essential for cutting out a shape from the interior of a stencil pattern.

Stencils can be any size or shape and there is no restriction to the number of pieces which can be used on the same article.

In stencil work use can be made of both negative and positive stencils (91 and 92), using different colours and achieving quite different results.

Soak the whole of the stencil in gum solution before placing it on the workpiece. It is best to use pointed tweezers to handle stencils as you must not fingermark the metal or the enamelled surface. Small pieces can be dipped in a dish of gum solution. The best way of soaking the bigger stencils is to spray them. To manœuvre the stencil, or stencil pieces, into position, dip a painting brush into gum solution, sat-urate the stencils, and you can then slide them where you want them (93). Remove excess solution and smooth down the edges with a dry brush.

After you have positioned the stencil pieces, give them a light spray with gum solution using low pressure. They will then be damp enough to stay in position when you start working with higher pressure.

The next step is to apply enamel by dusting, thoroughly covering the whole area outside the stencils. You cannot avoid dusting enamel on the stencils, however, especially when they are small (94). Do not try to remove the stencils when they are wet otherwise the gum solution will flow, taking with it some of the enamel to the areas which you wish to remain bare. But do not wait until the enamel is quite dry or the paper will shrink and curl, and when you move it the enamel will break away. The best time to remove the stencils is when the gum solution is half-dry. Also, the outline of the stencil under the enamel shows up best when the solution is in this state. Remove the stencil with pointed tweezers (95). If the area which you are covering with stencil is at the edge of the workpiece, leave a small overlap of stencil paper to make removal easy. If you do this, you will not need tweezers; lift the over-hanging portion with your fingers and pull slowly inwards.

When you have a number of small stencils, count them as you put them in position so that you can be sure that you have removed them all—rather like a surgeon counting his instruments. Complete the drying of the article before firing it. After the first firing, more colours, with or without stencils, can be applied in subsequent firings.

Leave stencils to dry—dry them flat if you want to re-use them—and then brush off the enamel with your fingers and save it for counter-enamelling. Enamel saved in this way should be washed through once in distilled water to remove excess gum.

(96 and 97)

*90* Cutting out the stencils

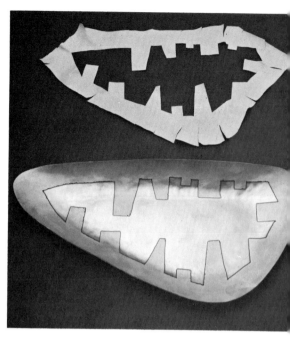

*91* The negative of the stencil is used to form the pattern in the first firing

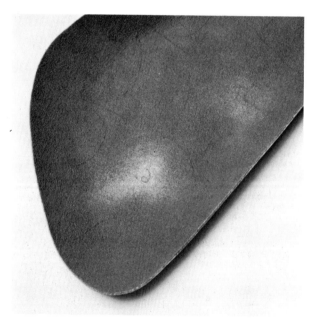

*94* Stencils dusted over with enamel. Only the edge of the large stencil piece need be covered with enamel.

*95* Removing the stencils in a half-dry state by picking up the edge with pointed tweezers

Soaking the positive stencils one by one before placing them on the workpiece

93 To adjust the position, they are flooded with gum solution and moved with pointed tweezers

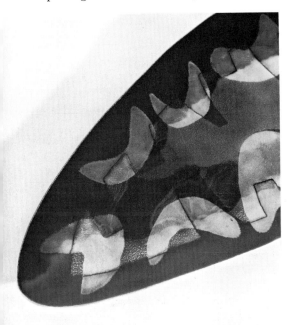

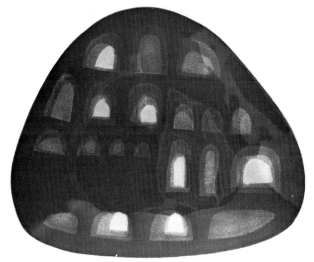

Detail of the finished piece. Although only two colours were used, a variation of shades was obtained as the colours were on copper, flux, and another transparent enamel

97 The use of positive stencils of the same shape but diminishing in size in each application

55

98 The design has been drawn and dusting is taking place
Note how the enamel sticks to the parts where glycerine has been applied

99 Enamel clings to the glycerined surface and the excess powder can be tipped off

## Glycerine

By using glycerine you can develop colour design in both line and area. It is a technique which is used only on pre-enamelled parts.

Draw your design with an ordinary pen using glycerine which has been coloured; it is necessary to add colour because gylcerine is clear and you cannot otherwise see properly what you are drawing. Dilute the glycerine with water. Undiluted glycerine is too thick to draw with; too much water will make it run and spread out of control. Add the water drop by drop, shaking the bottle well, until the solution flows easily from the nib. Put in a few drops of food colouring—bright red shows up well and also burns out in firing without leaving any trace.

Besides drawing your design you can paint areas in with a brush but be careful to wash the brush after use with soap or detergent in warm water to get rid of the glycerine. Dust the article with enamel (98) which will stick to the glycerine; shake off the loose enamel (99) and finally blow it gently. Place the workpiece in the kiln for two seconds, leaving the kiln door open. When it comes out you will notice that smoke rises from it. Leave it to cool and place it in the kiln again with the door open for another couple of seconds. Continue this procedure until no smoke is given off when it is brought out of the kiln. Then give the workpiece a short firing, watching it through the kiln peep-hole and remove it as soon as the enamel becomes smooth and shiny.

You can develop your design in further firings. Sometimes it will be found that the lines do not have sufficient body, in which case you can go over these again with glycerine, dust on more enamel, smoke, and re-fire.

This technique is suitable for opaque colours; however, with transparent enamels the glycerine tends to cloud them.

Very fine lines can be produced by using a drawing pen and dusting with 200 mesh enamel through a 200 mesh sieve.

(*100*)

*o* Finished piece. The last firing was a short overfire to allow the base colour to soften the lines of the pattern

## Inlay

Inlay is another technique with which you can create design in both line and area.

Your design should be developed, preferably in colour, on paper and then traced onto thin tracing paper. Place a sheet of carbon paper between the tracing paper and the copper surface and by going over your design with a pencil transfer it to the copper surface, which has been previously prepared as for enamelling in the usual way. The carbon lines on the copper surface should be scratched very finely into the copper with a needle point or a steel scriber (*101*). To get rid of all traces of carbon, thoroughly clean the copper surface with soap or detergent and a stiff brush, rinse it, dry it, and play the blow torch over it.

Now develop your design in one of two ways. Either work directly on the copper, or enamel the surface with a medium-fusing flux through which the lines you have scratched in the copper will be visible. Select your colours and apply them in the areas you have decided upon as described in the section of this chapter which deals with the technique of inlaying. If you find when doing the inlaying that you have excess moisture, this can be drained off by touching the edge of the soaked enamel on the workpiece with neatly cut blotting paper (*102*). A fine sable brush is very useful for straightening the edges and persuading the enamel into position (*103*). Before you start on a fresh colour, clean your tools so as not to mix colours. If the enamel which has already been inlaid has dried when you start on the next part, it will tend to absorb moisture from the new enamel and make spreading impossible, so it may be necessary to wet the original part. Dip your brush into distilled water to avoid an accumulation of gum in the enamel. Touch the edge and you will see it spread through the dry part. With a large article you can use a sprayer for this (*104* and *105*).

Your design can be developed in one or more firings. Remember that if you are working direct onto copper you must observe the rules for thickness of coating, but if you are working over a coat of flux, the colours can be of varying intensity and texture. For example, you can give them a stippled appearance.

## Combination of Techniques

You can, of course, use a combination of the application techniques described in this chapter on a single workpiece, or, indeed, invent your own techniques. The techniques referred to are known to produce satisfactory results, but there is no reason why the enameller should restrict himself to these. Enamelling is an art of almost unlimited scope and provides ample opportunity for creative improvisation.

(*106* and *107*)

57

*101* After the design is traced, it is scratched in with a scriber

*102* Excess moisture has to be removed to keep a sharp line at the edges where different colours meet

*103* Straightening an edge with a brush

104 Ready for firing

105 The finished piece. Inlaying an entire surface is a time-consuming task. This method was used by craftsmen centuries ago

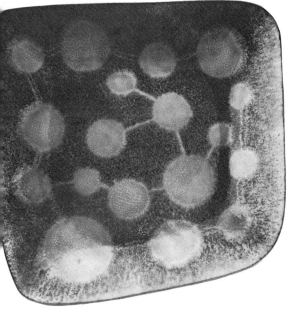

106 Stencil combined with sgraffito and over-firing

59

107 A combination of stencil, sgraffito, glycerine, and silver foil

# 5 Textures and Special Effects

## Threads

Threads of enamel can be bought, but as they are quite expensive you will probably wish to know how to make them yourself. If you do, you will find the task quite absorbing, and you will, moreover, be able to provide yourself with a wide range of colours at very small cost. You can also make the threads in a variety of thicknesses and lengths. Thread-making is a good way to use up left-over, roughly ground, and lump enamels.

You need an ordinary fireclay crucible (about size 4), such as is used in the silversmithing and jewellery trades; a pair of tongs for handling the crucible; a 15 mm ($\frac{1}{2}$ in.) steel rod about 176 cm (30 in.) long (this can be obtained from an ironmonger or builders' merchant); a hammer and a pair of pliers.

Thread making is a job for two pairs of hands, and, as it involves drawing molten glass through the air, it should be done by young people only when they are under the continual and strict supervision of an adult. It needs considerable preparation, so it is best to make at one session all the threads you are going to need for some time.

Fill the crucible with either lump or ground enamel and place it on a small firing rack. Put it in a cold kiln and heat to enamelling temperature. You can heat up together as many crucibles as can be accommodated in your kiln in a single row (*108*).

When the enamel is in a molten state at normal firing temperature, lift the rack bearing the crucible from the kiln on a firing fork. Then transfer the crucible with tongs to an asbestos mat placed on a heat-resistant surface such as marble, steel or asbestos sheet. One person holds the crucible firmly on the asbestos mat with the tongs (*109*). The other dips the steel rod, the tip of which has been heated to red in the flame of a blow torch, into the crucible

*108* Ready for thread-making. A row of crucibles filled with enamels on racks in the cold kiln

*109* The crucible is gripped tightly while the rod is dipped and the molten enamel drawn

and draws a strand of molten glass upwards, and, by walking away from the crucible for a distance of 1·8–2 m (6–7 ft), sideways. This stage is rather like withdrawing a spoon from a tin of treacle or a jar of honey. As the thread of molten glass is drawn through the air it cools and solidifies.

The thread is severed with pliers, first at the rod end and then at the crucible. You will find it necessary to draw the thread away from the rod with the pliers to allow the enamel to cool before it is brittle enough to cut. The thread may now be allowed to fall to the floor as by this time it is cool.

One heating will give you only about three or four draws, as the enamel soon becomes glutinous and unworkable and forms into a ribbon instead of a thread. Fill up the crucible again if you want to draw more threads of the same colour and replace it in the kiln. If the kiln is at firing temperature, you can now remove the other crucibles and start thread-making from these. In the meantime the enamel in the crucible you have just replenished will be reaching the correct molten stage, so in due course you can start making threads of your first colour again.

Let the crucibles cool off on an asbestos surface in a draught-free spot.

When the time comes to free the knob of enamel from the end of the rod before starting on another colour, plunge the end of the rod into a basin of cold water. Give it a few taps with a hammer and these will free the rod of the knob of enamel (*110*). Dry this enamel for re-use in the crucible or keep it for use as 'chunks'.

With a little experience you will find that you can make threads to your precise requirements. The thickness can be varied by the speed with which the rod is drawn away from the crucible; quick drawing produces a fine thread and slow drawing a thick one. Soft enamels are more fluid and when you use these, the thread can be allowed to fall back on itself to produce naturally attractive 'squiggly' patterns (*111* and *112*). You can make threads with any colour

you choose, or with mixed colours. If you melt different colours together in the same crucible, you can make multi-coloured threads containing streaks of different colours.

When the enamel threads have cooled, break them into lengths, and as the strands are brittle store in rigid containers.

The enamel which forms itself into a blob on the end of the rod, unlike that in the thread, retains its heat, so be most careful how you handle the rod. Care, of course, is also necessary in gripping the crucible firmly as this contains the hottest enamel. Another point to watch is that the thread as it is spun should not be allowed to touch anybody or anything, because although it is cooling in the air, it is still hot enough to burn. If it begins to sag, walk away with the rod more quickly. When producing 'squiggles', work over a heat resistant surface, such as that on which the crucible is placed. The person holding the crucible should wear heat-resistant gloves.

*110* Dipping the heated rod in cold water cracks the enamel blob and a light tap with a hammer frees it from the rod

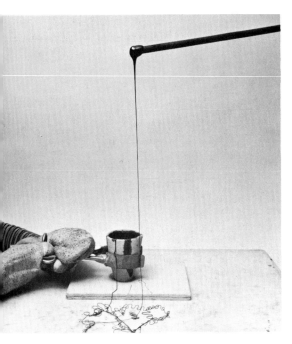

1 Soft enamels make 'squiggles' when taken out of the kiln but towards the end of the drawing straight threads can be made

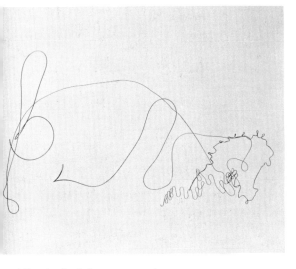

2 'Squiggles' form natural patterns

## How to use threads

Threads can be used either on a pre-enamelled surface or direct on copper which has been cleaned for enamelling.

You can use straight or curved pieces of one or more colours on the same workpiece. To bend threads, hold them over the flame of a gas torch; the enamel softens and enables them to be bent (*113*). You can also join pieces by fusing them together with a torch, or arrange threads in a pattern on an asbestos sheet and join them together by applying heat where they overlap (*114* and *115*).

Brush gum solution on the copper or pre-enamelled surface and place the threads on the article. If you are working direct on copper, select thicker threads. You will find a spatula handy to lift a joined pattern of threads on to a workpiece. Patterns which have been made up on a flat sheet of asbestos will not conform with concave or convex workpieces but when the enamel softens in the heat of the kiln the unsupported parts will sag and take up the shape of the article (*116*). You will find that by giving a short firing the threads can be made to stand out, while with a normal firing they will tend to flatten. There is a danger, however, that if they stand out too much the threads may crack and pop off in the cooling process or afterwards. Short broken pieces scattered at random over a surface give an interesting texture (*117*, *118*, *119* and *120*).

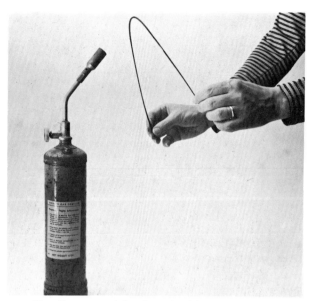

113 Bending a straight thread with the aid of a torch. The flame should be low, otherwise it will burn through the enamel

114 A very slight flame will soften the threads so that they fuse together

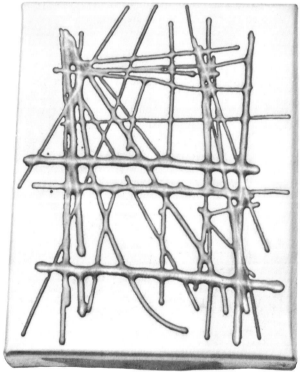

117 Threads fused on copper

118 A finished piece. The threads were of hard enamel to achieve sharply defined lines

*115* The pattern is very fragile at this stage

*116* The pattern laid across the pre-enamelled piece and ready to be fired

*119* Finished piece. The threads were of soft enamels to give a blurred effect

*120* Different coloured 'squiggles' fused in an opaque background

## Chunks

Chunks of enamel fired on to a surface produce pools of brilliant colour and also provide a variation in texture.

Brush gum on the pre-enamelled copper and place the chunks in position with tweezers (*121*). Watch the firing through the peephole in the kiln and remove the workpiece when the effect you want has been achieved. The longer the firing, the more the chunks of enamel will spread and flatten. The shorter the firing, the more the chunks will stand out, but do not allow them to protrude too much otherwise they may crack and pop off.

Chunks are not suitable for the steeply sloping sides of articles as they roll down before they have time to fuse. They combine well with threads (*122*).

Balls of enamel can be bought and these give small circular pools of colour of a more sharply defined pattern than chunks (*123* and *124*).

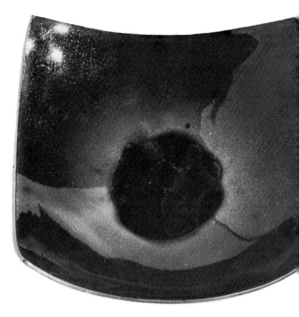

*122* Finished piece. Chunks produce brilliant pools of colour

*121* If you want to add more colour to your first coat, dust on the colour and then carefully place the chunks

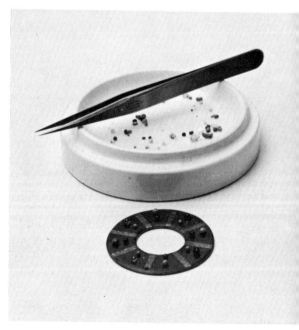

*123* Placing the balls. Pointed tweezers come in handy again. Use a drop of gum solution to prevent the balls rolling, and move the piece carefully before firing

*124* Finished piece. Balls give regular circular pools of colour

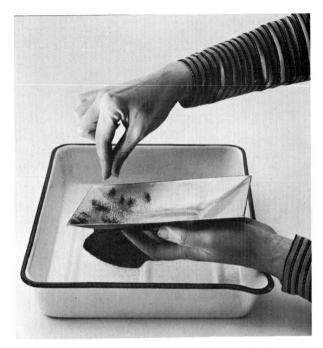

*125* Coarse enamels being applied with the fingers more or less at random

## Coarse

Coarse enamels, that is, those which will not pass through an 80 mesh sieve, can be put to good use by applying them to a pre-enamelled surface.

Spray the surface with gum solution and apply the large grains of enamel by sprinkling them with your fingers (*125*) or letting them fall from a small metal shute which can quite easily be made at home (*126*, *127*, and *128*). The enamel can be sprinkled at random or in a pattern. With both methods the result will be soft rather than sharply defined outlines. A single colour or a combination of colours can be used. To give a speckled look, coarse enamel can be sifted on a pre-fired coat with a sifter having a large mesh.

You will find that transparent coarse enamels produce lovely brilliant colours, particularly when overfired (*129* and *130*).

*126* Patterns are easier to execute with a metal shute

*127* More colours are added

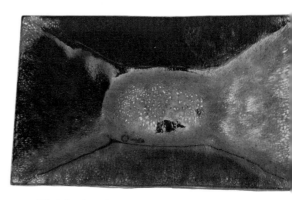

*129* Finished piece. The intermingling of colours was helped by a final overfiring

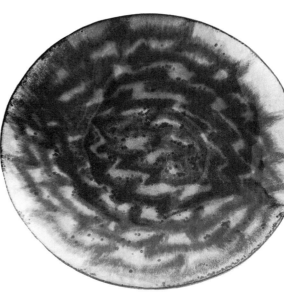

*130* Finished piece, again overfired in a final firing

*128* Ready for firing. Note that the coarse enamels are applied over a base coat

*The Frog Prince* – after Walter Crane
Enamel on steel 27 cm x 40 cm (11 in x 16 in)
by Richard Casey

*King and Queen* Painted enamel on gold foil by
Valerie Bexley

*Top* Champlevé enamel reliquary, made at
Limoges in the thirteenth century

*Bottom left* Painted enamel, Italian, late fifteenth
century

*Bottom right* Cloisonné enamel box made in China
not earlier than the Chien-lung period 1736–1796

British Museum London

## Fine

Enamels which in the grading process pass through a 200 mesh sieve can be mixed with gum solution and flooded on a pre-enamelled surface with a fine sable brush (*131*). This technique is similar to inlaying but much quicker as the fine enamels can be picked up and spread by brush. Comparatively fine detail, and even shading, can be achieved, especially when the enamel is scraped and worked after it has been allowed to dry. A normal firing is given but you should be careful not to be too liberal with the gum solution or the transparent enamels will become foggy (*132*).

*Very finely ground* enamels can be applied to a pre-enamelled surface with a brush rather as if you were painting in oils. To grind the enamels to a sufficiently fine powder—about 400 mesh or finer—you will need an agate pestle and mortar, which is rather expensive.

After washing and drying the enamel, mix a small quantity of the fine powder on a ground glass slab with oil of lavender using a palette knife or glass muller. Add the oil drop by drop until you have a paste. Add a drop or two of fat oil to give the paste body and make it easier to spread. An oily edge when you paint on the enamelled surface means there is too much oil in the mixture.

Choose the brush and the method of application to suit the work. As the coat produced by these very finely ground enamels is very dense, it can be thinner than one applied by dusting, but do not spread it too thinly or the colour will burn out in firing.

*Special painting colours* can be bought from enamel suppliers. The advantage of these colours, apart from the fact that they are ready-ground, is that they are very highly pigmented and the thinnest application will leave an intense colour after firing. The finest shading, akin to that obtained with water colours, can be produced. Painting colours are mixed with oil of lavender and fat oil, in the same way as the very finely ground enamels mentioned above, and brushed on. The appearance of the enamelled surface when wet is almost the same as after firing. The coat should be thin, just sufficient to give the required colour intensity. These colours are so finely ground that they can be mixed with each other to form additional colours.

When working with oils, clean the brushes and mixing slabs with a proprietary brush cleaning liquid.

After you have completed your painting with oils, whether using finely ground enamels or painting colours, place the workpiece in a hot kiln for about two seconds, leaving the door open. When you take it out you will see smoke rising from it. Let it cool for a few minutes and repeat this process until the article gives off no more smoke. The firing should be short. Watch the workpiece through the kiln peephole and take it from the kiln when the surface becomes shiny. Add to the design, or strengthen the colours if necessary, in subsequent firings.

(See also Section 6, *Painting Enamels*.)

*131* Flooding on fine enamel

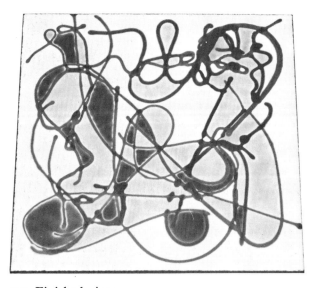

*132* Finished piece

## Silver and gold foil

A transparent colour naturally looks different on different metals. By introducing silver or gold foil into enamelling on copper you can produce the appearance of silver and gold without the expense. Before starting with silver and gold foil, consult your transparent colour samples (see page 90) so you use the foils and colours to their best mutual advantage.

Silver and gold foil must be enclosed between coats of enamel. Special foil for enamelling can be bought. Leaf is not suitable; it is too thin and burns out. Foil is supplied by the manufacturers in books between sheets of paper which protect it. Leave it there until you use it. Treat these foils as a metal surface cleaned and ready for enamelling; never touch them with your fingers.

Silver and gold foil is applied over a pre-fired coat of enamel which has been prepared in the usual way for the next coat of enamel. Remove the foil from the book together with the two protecting sheets of paper. Place it on a cardboard backing and punch holes through the paper and the foil with a needle point (*133*). The purpose of this is to allow air and gases trapped between foil and enamel to escape. Draw or trace your design on the paper covering, taking care to position it so that you avoid excessive offcuts. Clip together the foil and sheets of paper in order to prevent them slipping and with scissors cut along the pencil line (*134*). A stencil knife is not suitable for this, as it will cause the foil to ruckle up between the sheets.

Brush gum solution on the workpiece and then, without dipping the brush into the solution again, pick up the foil with the brush (*135*). Place it on the enamel with the same side uppermost as when you did the piercing. Using tweezers to help you position the foil where you want it, smooth the foil down with the brush, keeping it in place with tweezers. Do the smoothing from the centre outwards so you squeeze out any pockets of air or excess gum which may be trapped.

Dry the workpiece and give it a short firing. The foil will sink into the enamel, giving a slightly rippled effect when fired. If you are after a smooth effect, however, working rapidly, press down the foil with a palette knife as soon as you take it out of the kiln while the enamel is still soft (*136*). Taking care not to touch the foil with your fingers, file the fire-scale from the bare edges of the workpiece. Do not pickle. Polish the foil surface with a glass fibre brush (*137*) and play the flame of the blow torch over the whole surface. Dust on, or apply by inlay, transparent enamel and give a normal firing. After firing, you can put on another coat and fire again. Observe the rules for transparent coats—instead of one thick, use two thin.

Again, this is a technique which can be combined with others. For example, chunks of enamel over foil give a brilliant sparkle (*138*).

Punching holes in silver foil

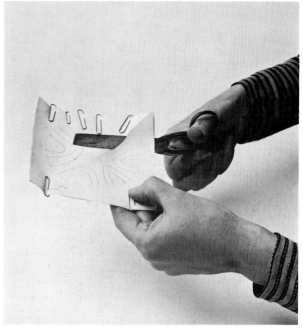

*134* The traced pattern is cut

*135* Positioning the foil with a brush

*137* Polishing with a glass fibre brush after firing

*136* If a rippled effect is not desired, foil can be smoothed out

*138* Finished piece. Transparent enamel over foil gives a real sparkle. Blues go well over silver, and rubies over silver give the impression of a rich gold

# Swirling

All the techniques described so far create patterns before firing. By swirling you can manipulate molten enamel on a workpiece while in the kiln. It is a technique particularly suitable when chunks or threads, or both, are applied on a base coat of enamel.

Arrange the chunks and threads in areas and place the workpiece in the kiln for the length of a normal firing. The surface of the enamel is then lightly swirled or disturbed in its molten state with a special tool which has a wooden handle in which is inserted a pointed wire bent at the end. Do not use too much pressure when wielding the swirler.

Pre-heat the swirler either in a blow torch or in the kiln before you dip it into the molten enamel. After swirling leave the workpiece in the kiln for a few seconds to allow the enamel to flatten out.

If you are working with a blow torch or a hotplate kiln, the swirling can be done without a screen. With a kiln, however, a steel screen is essential to protect your face from the heat, and also to prevent the heat escaping from the kiln.

When swirling you may make an open furrow in the enamel where the enamel is thin or has been cooled down too much. If this happens, clean the exposed metal with acid to remove firescale, patch up the hole using the inlay technique with the appropriate colour, and re-fire the piece. If an air bubble should be trapped in the enamel, break it with a scriber, fill up with enamel, and re-fire.

After using the swirler, let it cool and with pliers crack off the blob of enamel which has formed on the end.

Sharpen the swirler from time to time, but not to a needle point.

(139–149).

139 Swirling on a hotplate kiln

140 You have good control over your swirling in this method of firing

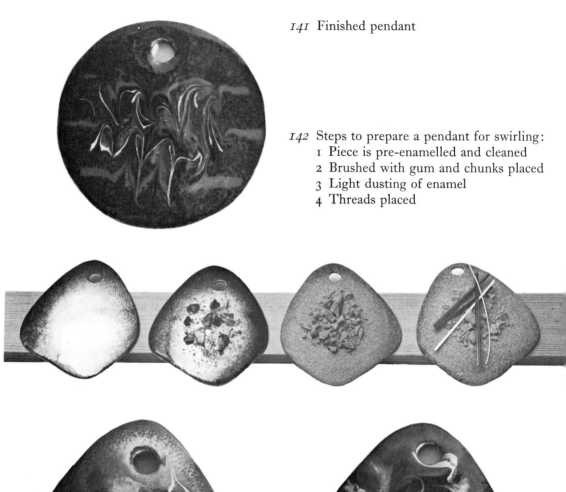

*141* Finished pendant

*142* Steps to prepare a pendant for swirling:
1 Piece is pre-enamelled and cleaned
2 Brushed with gum and chunks placed
3 Light dusting of enamel
4 Threads placed

*143 and 144* The chunks and threads were arranged
identically, but you can achieve completely
different results in swirling

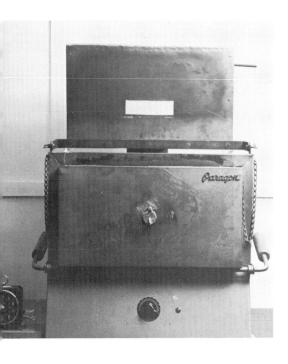

145 A face shield is necessary for swirling in a kiln. Two brackets are fixed to the sides of the kiln and the shield is fixed with bolts and wing nuts

Open the kiln door with one hand and lower the shield with the other, quickly to prevent loss of heat

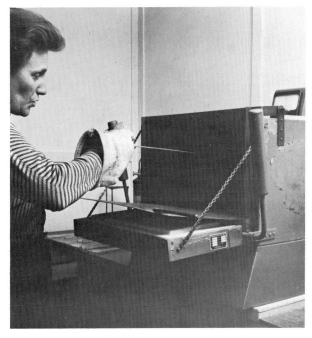

147 The swirling tool is inserted and the swirling done through a slot in the shield

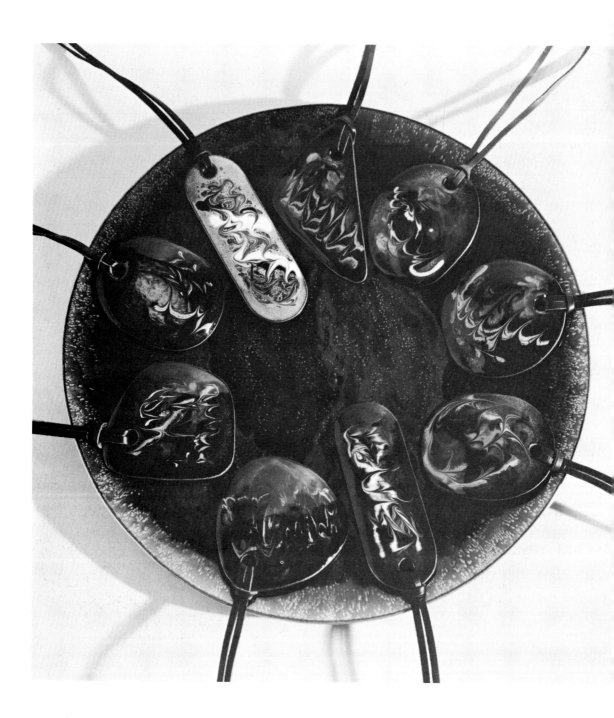

The more experience you have of overfiring, the more control you will have over the final result. There is always an element of chance involved but in time you will acquire a high degree of control.

Gently sloping surfaces give the enamel a chance to flow. Steep sides, however, cause the enamel to run down and form a pool on the bottom. If some of the colours need strengthening, you may apply a further coat, always giving a heavier application on the higher parts. You can give the piece a final firing or overfire again.

(*150* to *154*)

*149* Bigger pieces can be swirled in a kiln

## Overfiring

With normal firing, the temperature is sufficient to fuse the enamel to the metal without disturbing the intended pattern. If you want your colours to run and flow into each other, you should overfire the piece.

Overfiring means firing enamels not for a longer time but at a higher temperature—approximately 100°C (212°F) above the firing temperature recommended by the manufacturer for the enamels you are using. For overfiring you apply colours on a pre-enamelled surface. Almost any transparent or opaque colour will serve as a base coat but coarse transparent enamels over a hard flux base, or transparent or opaque enamels dusted on a soft white base, give good results. Some of the opaque colours (turquoise and mauve, for example) will become transparent when over-fired. (Avoid opaque reds or oranges as they lose their colour in the higher firing temperatures.)

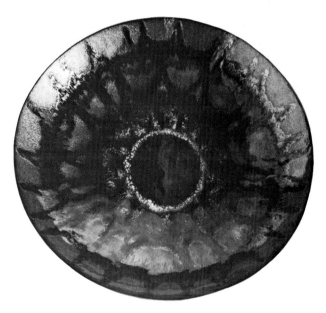

*150* A bowl. Coarse enamel was applied with a shute and overfired

*151* A large bowl. Coarse enamel was applied
   with a shute and overfired

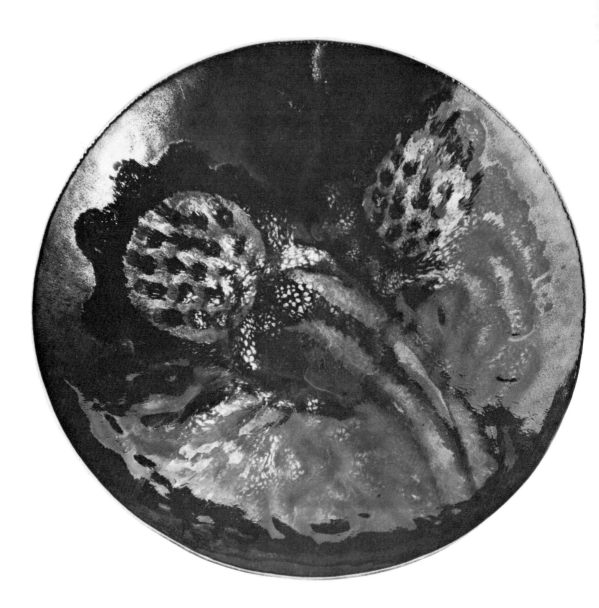

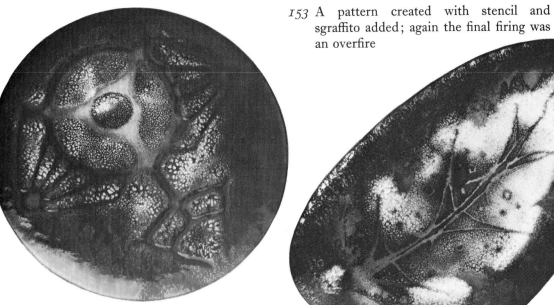

153 A pattern created with stencil and sgraffito added; again the final firing was an overfire

152 This pattern was made by stencilling; the final firing was an overfire

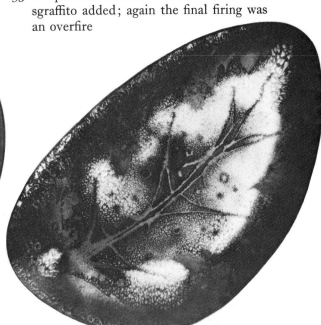

154
1 Coarse enamels, finger applied
2 Coarse enamels, shute applied
3 Deep etching, recesses filled with inlay technique
All three pieces were overfired

# 6 Control of Design

## Crackle Enamels

Self-cracking enamels are most effective on the curving surfaces of bowls and ashtrays. Applied over a prefired base of enamel, they split and crack during firing, leaving a fine network of lines through which the base enamel shows.

They come in the form of a very fine powder which is mixed with water to a paste. This paste must not be too thin otherwise it will run down into the centre of the bowl without adhering to the sides. Apply the paste as evenly as possible over the chosen area, not too thickly or the cracking will not be so effective. The whole inside of the bowl can be covered, or the cracking enamel can be applied to certain areas between the other colours. In the U.S.A. and Canada crackle enamels come in liquid form which should be well stirred before use; they should not be allowed to dry out and if necessary can be thinned with water. The liquid can be brushed or sprayed on or the piece can be immersed in the liquid and the excess shaken off.

Foil used under transparent enamels makes an extremely effective base, as the foil sparkles between the lines of the crackle enamel. There is a variety of effects available. For instance, transparent colours can be overlaid on top of the crackle enamel after it has been fired. Make sure that the enamel is dry before firing by placing the bowl on top of the kiln for about ten minutes, then fire as usual. It may be found that different colours have slightly different ways of cracking. The main lines of cracking can be controlled by drawing a needle through the powder just before it is completely dry. The enamel then cracks along the lines made.

Crackle enamels make a good counter enamel or base coat as they do not crack if fired directly on copper.

(155 and 156)

155 The liquid crackle enamel is applied to the inside of the bowl, using a brush

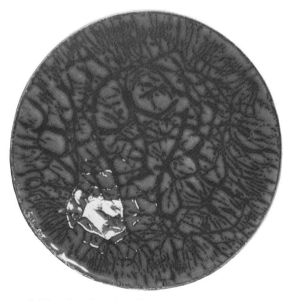

156 The finished bowl, the base coat showing quite clearly between the cracks

## Inset of Wire Shapes

The insetting of designs by wire shapes into enamel is a fairly easy technique. The only problems are those which arise in the bending of the wire and these can be overcome by practice with odd pieces of wire. Before use, the wire must be annealed to make it pliable. Place the wire on a mesh and put it in the kiln. When it glows red take it out and plunge it into cold water. To smooth it draw it backwards and forwards over the edge of a table until all the kinks are gone. If you are using copper wire, polish it with steel wool to remove the firescale. The main tools necessary for bending the wire are a pair of jeweller's pliers and tweezers but you can bend the wire to obtain various curves and shapes—the ends of round-barrelled pens and pencils are useful for making circles. The design of the wire can be spontaneous, make it up as you go along, or you can first draw the design on paper and, holding the wire against it, carefully bend the wire to follow the drawn line.

It is best to fuse the wire into a base of transparent enamel as this melts more quickly than opaque enamel and the wire fuses to it without a layer of oxide having formed on it which would cause it to crack off afterwards. If you want to use opaque enamel as a base, fire a layer of flux over the top. The wire must be free of all grease, as this may prevent fusion, so brush the piece all over with thinners. Coat the enamel base with a generous layer of gum and position the wire shape in it, then allow the piece to dry before it is fired.

The temperature of the kiln must be about 850°C (1562°F) (this allows for a drop in temperature when the door is opened). Watch the piece during the firing and when it glows red withdraw it as you do not want the wire to sink too far into the enamel or to be covered. The wire may not have been lying completely flush with the enamel surface. If, when you bring the piece from the kiln a part of the wire is protruding, carefully push it down with a pair of tweezers or the sgraffito tool. When the piece is cool, clean the wires and edges of firescale with steel wool. If a proper shine is desired then give a final rub with metal polish.

The same firing procedure should be followed for inset of flat metal shapes. (*157* and *158*)

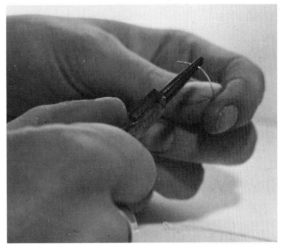

*157* Bending the wires

*158* The finished piece. Here the wires were fused into a base of pale amber transparent enamel. Work of Diana Hull

## Cloisonné

This is one of the oldest techniques, one which has been practised for several hundreds of years.

The enamel is used as in inlay work, except that instead of being laid straight onto the metal base it is placed into 'cloisons'—shapes made by wires which have been fused into a layer of enamel (*159*). The wires most frequently used are copper and silver; gold can be used as well. Years ago craftsmen soldered the wires straight to the metal base, but now it is customary for the wires which have been bent to form the design to be gummed to the enamel surface (as for insetting) and fused until they are partly embedded in the enamel, so that a tiny ledge is left which encompasses the enamels, and prevents those in one division from flowing into another.

When packing the cloisons, make sure that the level of the enamel powder is slightly higher than the wires and that it is thoroughly dried before firing (*160*). After firing, the enamel becomes compacted and sinks below the level of the wires. You may need to pack and fire two or three times before the level of the fused enamel is just above that of the wires.

Remember to clean copper wires of firescale between each firing otherwise bits of firescale will flake off the wires onto the enamel; removing these bits is a laborious task.

When the level of the enamel is above the wires it is time to start stoning it down. The aim is to make the whole surface of the piece perfectly smooth. Wires and enamel should be level, but this is impossible to achieve simply by firing the enamel to a perfectly flat surface—the slightly lower fusing enamels would overflow the wires, the colours immediately becoming confused with each other.

Stone the piece under running water with a carborundum stone, making sure that all particles of dirt are washed away to prevent them contaminating the enamel. When the enamel surface—now matt—is level with the

*159* First the wires are fused into a base of flux

*160* The cloisons are filled with enamel, using the wet-packing method

wires, examine the piece closely. If there are shallow areas still shiny, these must be re-packed with enamel and fired, and the piece stoned again. Repeat this process of repacking, firing and stoning until the whole surface is level. It is best to use a slightly finer carborundum stone each time.

Any small specks of dirt or firescale which have entered the enamel must be carefully drilled out, repacked and fired.

When the surface is smooth and level, either flash fire or hand polish using felt and cerium oxide, both wet. Flash firing is firing at a fairly high temperature very quickly to bring back the gloss. Hand polishing should be employed where a less glossy finish is desired. (*161* and *162*)

*161* The piece was wet-packed twice with enamel (firing in between) to bring the enamel slightly above the level of the wires, then stoned flat under running water, dried completely, and flash fired, to bring back the gloss. Work of Judy Williams

*162* Cloisonné enamel cross by Diana Hull. Silver and gold foils were fused to the flux in the cloisons. Transparent enamels were wet-packed over the top

83

*163 Angel Musician*, champlevé, by Diana Hull   See also page 18

## Painting Enamels

Painting enamels are very finely ground and highly pigmented enamels which are mixed to a smooth-flowing consistency with oil or gum and then applied either by painting or spraying onto a base coat of prefired enamel. A variety of oils is available—white spirit, fat oil thinned with a little lavender oil, tramil and 'painting medium' are obtainable from craft shops. The colours are usually applied to a white base, although other pale colours and flux can be used to good effect as long as they are not hard-firing enamels. Painting enamels should be fired at temperatures in the region of 800°C (1472°F) or they will fade. Hard-firing enamel fired at this temperature will split and cause cracks in the design. The powders should be mixed thoroughly with the painting media using a palette knife (*164*) and ensuring that

*164* Mixing the painting enamels with the medium using a palette knife

all the small lumps of powder are completely dissolved in the oil. Unless this is done, a smooth, shining surface cannot be achieved when the piece is fired. The consistency should be neither so thin that the colours turn out too pale nor so thick that the mixture is unmanageable. A little experimenting is necessary to find the right proportions of powder and oil.

Unlike ordinary enamel powders, the colours are very similar before and after firing, though slightly different hues can be obtained with blues and greens by extending firing time fractionally. The colours can be mixed to form different shades, but in this case the above rule no longer holds, and testing is necessary when mixing a new colour, as the difference between the mixture on your palette and the fired result can be striking. Pale colours, specially those with pink and red in them, become a murky grey and skin tones become too yellow or too brown. This is due to the strength of some colours being greater than others and overpowering them. So testing new shades before applying them to the workpiece is vital.

One of the hardest, and often most desired, effects to achieve when using painting enamels is the coverage of a large area evenly with one colour, or the gentle shading necessary for clouds, one colour subtly fading into another. A sable brush gives very good results; it is stiff enough to hold the paint without absorbing it too much and yet does not leave unwanted brush marks.

The drying of the enamel, whether just before firing or between painting adjacent colours, is important and can cause problems. If the piece is dried too quickly the heat can cause the colours to run, and some beautiful shading can become a sludgy mess. Do not dry the piece on top of the kiln as this gives off too much heat; a table lamp positioned over the piece produces enough heat to dry the enamel without spoiling it. Once dry to the touch, the final drying process can be carried out. First place the piece on top of the kiln and then

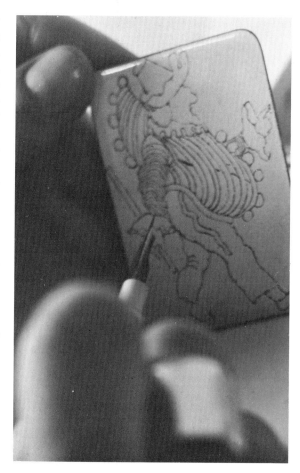

165 Outlining a piece before colouring using a mapping pen. The mixture of painting enamel and medium must not be too thick or it will not flow through the nib

insert it inside for a few moments until the oils are completely dried out.

As they fire so quickly the firing of these colours is not easy and requires practice for real proficiency. As little as half a minute is all some colours need. As mentioned earlier, only soft to medium firing enamels should be used as a base, so fire the painting enamels at the same temperature as that needed for firing the base you are using; fire any higher and the colours will fade, fire too low and the base enamel will crack, pulling the design apart. As the firing is quick, watch closely and withdraw the piece from the kiln as soon as a faint shine appears on the surface; if the colours are still a little matt insert the piece in the kiln again for a further few seconds.

A few colours do, however, take much longer to fire, notably blues and browns, and it is advisable to apply and fire these first. Reds and yellows fire more quickly than the other colours and tend to fade while the others are still fusing. The best way is to apply these colours when the rest of the piece has been fired and fire them last of all and very quickly; the other colours will not be affected.

(*165* and *166*)

*166* Painted enamel pendant. The outline was applied first, using a mapping pen, and fired. The various colours and shading were applied together and the piece fired a second time

*167 Jonah* by Valerie Bexley. Painted enamel on foil

## Lustres

Gold and silver overglaze lustres can look very beautiful. They have a metallic finish which is effective because it is not harsh and contrasts well with the enamel. Lustres can either form their own design or outline a design executed by the inlay method (black painting enamel is very good for this), or give additional touches to a painted enamel piece.

The lustres are obtained in liquid form and are applied to the fired enamel with a brush, pen or spray. Care must be taken not to apply them too thinly or they turn out a watery purple colour, nor too thickly, in which case they remain black. If you accidentally smudge or smear the lustre, the area must be thoroughly cleaned. Wiping with a rag or tissue will not suffice as the smear will still appear when the piece is fired; it is probably best to clean the whole piece with thinners and then with soap and water to ensure that all traces of the lustre have been removed.

The lustres must be carefully dried before firing. To do this insert the piece in the kiln and withdraw it after a few seconds. There is an extremely strong smell as the oils quickly evaporate, and fumes can be seen rising from the piece. Repeat several times until the fumes stop rising from the surface. Firing is quite quick; watch closely and when the colour of the lustre appears leave the piece in the kiln for a further few seconds to make sure that complete fusing takes place.

If the lustre appears rather weak and purplish in colour, or small cracks have appeared in it due to slight overfiring, repaint the piece and fire again.

(*167* and *168*)

*168* Here a traditional colour scheme was followed, the robes painted in tones of blue and gold lustre for the background

# 7 Colour Samples

As the colours of both opaque and transparent enamels are different in lump and ground form, and when fired, the making of colour samples is essential.

## Opaque

Prepare a strip of copper (25 × 50 mm—1 in. × 2 in. is a practical size). Drill a hole at one end so that it can be hung up for easy reference. Counter-enamel it. Dust the colour over the whole metal surface and fire it. Prepare it for the next coat. Cover the top half with stencil paper (*169*), dust on a second coat of the same colour, and fire. You now have a sample which indicates how one and two coats of enamel will look. It is particularly important to know this with reds and oranges, which vary a great deal according to the number of coats. Label the back of the sample and write on it in ink the name of the manufacturer, the colour, and the manufacturer's code for the colour (*170*).

## Transparent

You will need samples of transparent colours over copper, over a coat of pre-fired white, and over silver foil, gold foil, and flux. You must know how transparent colours look on these different bases so as to be able to use them to their full advantage. You will find that colours of transparent enamels vary a great deal on different bases.

A good size for these colour samples is 75 × 25 mm (3 in. × 1 in.). Counter-enamel the strip. Divide the copper into six parts. Cover the bottom 25 mm with stencil paper. Dust a hard flux on the remainder. Fire the piece. Prepare it for the next coat. Cover the flux with stencil paper and also the bottom 13 mm. Dust on a hard white enamel between the stencil papers and flux. Fire it and prepare it for the next coat. Now all the strip is enamelled except the bottom 13 mm. Apply silver foil adjacent to the white enamel and gold foil immediately above that. Follow the same order in all the samples,

*169* When making samples, protect with a stencil the part which is not going to receive enamel

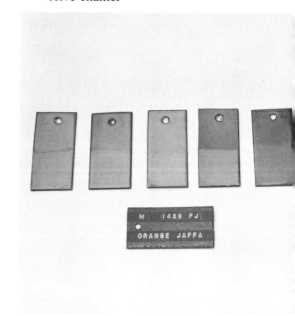

*170* Opaque samples. Most colours differ if applied in one or two coats

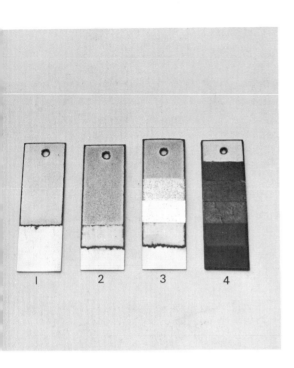

*171* Transparent colour samples in different stages of preparation

1 Top 50 mm (2 in.) covered with hard flux, lower 25 mm (1 in.) copper

2 13 mm ($\frac{1}{2}$ in.) hard white added adjacent to the flux.

3 13 mm silver foil above white and 13 mm gold foil above silver

4 Transparent colour except on top 13 mm

because with some colours you will not be able to tell which is gold foil and which is silver when they are covered with enamel. Fire the piece again. Then place stencil paper on the top 13 mm of the sample and dust colour on the rest of the strip. Fire once more (*171*). See drawing C.

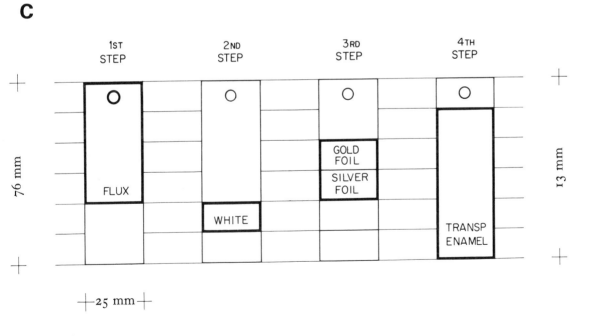

C

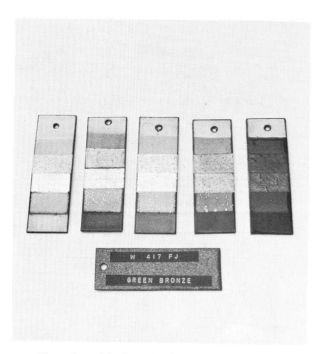

The strip of exposed flux at the top of the sample gives a good indication of the intensity of the colour in comparison with colourless flux.

Label as for opaque samples (*172* and *173*).

When samples have been finished, file around the edge to get rid of firescale.

As preparing samples involves quite a lot of work, it is best to do several at a time.

*172* Even in a black and white photograph it can be seen how different the same colour looks on copper, opaque white, silver foil, gold foil, and flux

*173* Well-arranged colour samples can make a gay array

92

# IV Finishing

After the final firing it is necessary to clean firescale from the copper which remains exposed. The parts requiring treatment are usually the edges and, where the counter-enamel does not cover the whole of the back, the base. It is not advisable to submerge the workpiece in an acid bath as some colours may be affected.

## Bases

One way to clean the bare copper on the base is to rub it with a piece of cotton wool soaked in nitric acid solution, holding the piece with tongs (*174*). With obstinate firescale, such as is obtained in over-firing, you may have to leave the plug of soaked cotton wool on the workpiece which has been placed face downwards. Take care that acid does not run over the sides. Move the cotton wool swab from time to time so that the whole surface is cleaned. After cleaning, rinse under running water, wash the surface well with detergent, dry and polish it with fine steel wool. This will not scratch the enamel.

If you have a polishing motor, you can rid exposed copper of firescale by using the same mop and cleaning compound as for normal cleaning preparatory to enamelling. Be careful not to polish the enamel as this dulls the surface.

## Edges

The edges of an enamelled article are cleaned by using a carborundum stone of medium grade. Dip the stone and article in water frequently as you work. Use round or flat stones according to the curve of the edge. Start stoning on the face of the article and draw the stone across the edge with outward strokes (*175*). This smoothes off the edge of the

enamel. Turn the article over and clean the outside edge in the same way (*176*). You can then work the stone along the copper edge from side to side to complete the job of cleaning the metal (*177*). Take care not to stone the enamel surface. Give the piece a good rinse in running water and dry it thoroughly. Polish the edges with steel wool until they feel smooth to the touch.

It is a good idea to cover the end of the stone with drafting tape to prevent the sharp edge accidentally scratching the enamel surface. If you do accidentally scratch a piece when finishing it, scrub it well to remove all traces of stone particles, prepare it for enamelling, apply a very light coat of enamel, and re-fire.

When stoning do not try to remove the dark edge of the enamel as this is characteristic of hand-made pieces and is part of their attraction.

## Lacquering

After a time copper exposed to the air will oxidise. A coating of colourless lacquer will protect the metal and prevent this happening.

## Findings

A wide variety of jewellery findings, such as earring clips and brooch pins, can be bought.

Clean the surface to which you are fixing the finding and also clean the finding itself. Rough up with carborundum stone the enamel where the finding is to be fixed, and scratch the bare metal of the finding or the workpiece with a scriber in order to give a key for the adhesive. Epoxy resin glues are very satisfactory for securing findings to enamelled articles.

Choose findings which have fairly broad surfaces at the joining area as they give a more satisfactory bond.

174 Final cleaning of the exposed copper base with cotton wool soaked in diluted acid

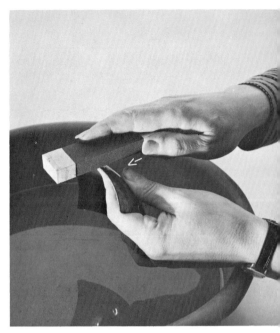

176 Similar outward strokes are used on the back edge

175 Smoothing the face edge with one-way outward strokes

177 Cleaning the copper edge; now you can work from side to side. Continue stoning until all traces of firescale have been removed

*178*

# V Tips for the Beginner

The proportion of opaque and transparent colours one uses is, of course, a matter of individual choice. The authors' consumption of enamels is roughly 30 per cent flux, 20 per cent opaque, and 50 per cent transparent.

The colours in enamel are provided by metal oxides, and some oxides are more expensive than others. Opaque reds and transparent rubies are among the dearest, but it would be a pity to dispense with them completely. Rubies over silver foil give one of the richest and most beautiful colours in enamelling.

Do not order too many colours to start with; build up your colour range gradually. It is suggested that initially you obtain the following range of transparent enamels: hard and medium flux, dark and light ruby, light, medium and dark blue, medium brown, orange or yellow, light and dark green, turquoise, and one shade of mauve or lilac. In opaques you may begin with hard and soft white, dark and light red, light, medium and dark blue, medium or dark brown, orange, yellow, light and dark green, turquoise, mauve or violet, light or medium grey and black.

Ground enamels go a long way; for example, 500 g (1 lb) of medium grade flux would cover two 30 cm (12 in.) bowls, two 25 cm (10 in.) plates, two 20 cm (8 in.) plates, four 15 cm (6 in.) bowls, one 15 cm plate, one 20 × 12 cm (8 × 5 in.) dish, one 10 cm (4 in.) bowl, one 12 × 10 cm plaque, and fifteen 65 mm (2½ in.) diameter pendants.

Orderliness and cleanliness are essential in enamelling. Your materials and equipment should be placed so that you know just where they are and reach for them automatically. Keep your workshop as clean and dust-free as you can. Clean your sifters, sieves, and bowls immediately after use. Foreign matter in your enamels and colour contamination lead to poor results.

The range of articles which can be enamelled is almost limitless (*178* and *179*). It covers objects for everyday use in the home, such as door knobs and fingerplates, coffee-table tops, trays, clocks; items to wear, such as cuff-links, buttons and pendants; and articles which are useful as well as decorative, such as vases, bowls, plates, ashtrays, cigarette and other boxes.

Finally, there are purely decorative pieces, murals and plaques, which will add interest to the walls of rooms, foyers, or even the external walls of buildings.

*179*

*180* Fruit dish, 35 cm (14 in.) diameter by
Richard Casey. Enamel on steel

# VI Brief Historical Notes

No one knows when the first enamelling was done. Jewels of the Mycenæan period were decorated with a blue vitreous substance, and six gold rings discovered in a Mycenæan tomb of the thirteenth century B.C. in Cyprus were decorated with cloisonné enamelling. There is doubt, however, whether the enamels in these early examples were applied in the form of powder or were merely fragments of coloured glass. The art of enamelling may have been inherited by the Greeks and spread by them to other parts of Europe. Bronze ornaments dating from the ninth century B.C. found at Koban in the Caucasus bear decoration in enamel.

During the Roman period enamelling was carried on in the old Celtic areas and in North Africa. Enamelled horse trappings have been found in many places in the British Isles and these are believed to date from the first century A.D. Examples are in the British Museum in London. The same museum also contains Saxon bronze ornaments decorated with enamel excavated from the Sutton Hoo Ship Burial shortly after the Second World War. The hanging bowls in this find are good examples of champlevé work.

The important period of Byzantine cloisonné enamelling began around the tenth century in Constantinople; cloisonné first appeared in Russia about the same date.

The champlevé process flourished some two centuries later in the valleys of the Rhine and the Meuse and at Limoges in central France. The Mosan and Rhenish enamels have green and yellow as dominant colours and the blues are paler than those used by the Limoges school. The best work from Limoges was done in the twelfth and thirteenth centuries.

A new technique, known as basse-taille, was developed in the thirteenth century in Italy. An example of French basse-taille work is the Royal Gold Cup in the British Museum. The translucent enamel is applied over sunk relief.

Early in the fifteenth century another technique, plique-à-jour, came into prominence. Few examples of this form of enamelling survive today, as, being fused into an open-work frame, these enamels were extremely fragile.

Enamel colours were painted on glass in Venice in the fifteenth century, and the Venetian and Limoges enamellers adopted this method on metal and created a new technique with their painted enamels. The work of early Limoges enamellers is well represented in the Louvre in Paris. Enamelling continued to flourish in the second half of the sixteenth century. One development of the painting technique was the carrying out of designs in grisaille on a dark blue or black ground.

Watch-cases were enamelled in the seventeenth century, and later snuff-boxes. In England, both Birmingham and Bilston became centres for painted enamel towards the middle of the eighteenth century, and étuis and other small objects made of copper were enamelled in fine detail and delicate colour. At first, the enamels were hand-painted. Transfer-painting was invented in Birmingham in 1751, and was brought to great perfection at Battersea between 1753 and 1756. The products of Birmingham, Bilston, and Wednesbury have often been wrongly described as *Battersea enamels* (*181*, *182* and *183*).

The world's museums contain many examples of Chinese enamels of the fifteenth century and later, cloisonné, champlevé and painted. In Japan, enamelling is said to date back to the seventh century. There was a resurgence of the craft in the nineteenth century and the realistic, minutely detailed designs in lineless cloisonné were popular.

Enamelling was revived in England by Alexander Fisher and others in the art nouveau period.

181 Etui, enamel on copper, mounted in gilt-metal, painted in colours and gold. Probably made at Birmingham or Bilston, about 1760

182 Trinket box, enamel on copper, mounted in gilt-metal, painted in colours and gold. Probably made at Birmingham, about 1755

Plaque, enamel on copper, in gilt-metal
frame, transfer-printed. *Les Amours
Pastorales*. Battersea or Birmingham,
about 1755

*181–3* In the collection of W. E. Benton

# Gauges for Sheet Copper

For specifying the thickness of copper sheet, Standard Wire Gauge (SWG) is used in the United Kingdom and American Wire Gauge (AWG), also known as Brown and Sharpe Wire Gauge (B&S), in the United States of America.

| SWG | mm | in. | B & S |
|---|---|---|---|
| 24 | 0·559 | 0·022 | 23 |
|  |  | 0·025 | 22 |
| 22 | 0·711 | 0·028 | 21 |
| 21 | 0·813 | 0·032 | 20 |
| 20 | 0·914 | 0·036 | 19 |
| 19 | 1·016 | 0·040 | 18 |
|  |  | 0·045 | 17 |
| 18 | 1·219 | 0·048 |  |
|  |  | 0·050 | 16 |
|  |  | 0·057 | 15 |
| 16 | 1·626 | 0·064 | 14 |
| 15 | 1·829 | 0·072 | 13 |
| 14 | 2·032 | 0·080 |  |

# Bibliography

*Enameling Principles and Practice* Kenneth F Bates The World Publishing Company, Cleveland

*Metal Enamelling* Polly Rothenburg Allen and Unwin, London

*Enameling for Beginners* Edward Winter Watson-Guptill, New York

*Enamel Art on Metals* Edward Winter Watson-Guptill, New York

*Enamel Painting Techniques* Edward Winter Elsevier Publishing Company, Barking, Essex

*Enameling on Metal* Oppi Untracht Chilton Company, Philadelphia

*The Craft of Enamelling* Kenneth Neville Mills and Boon, London

*The Art of Enamelling* Margaret Seeler Van Nostrand Reinhold, New York

*Metalwork and Enamelling* Herbert Maryon Constable, London

*The Preparation of Precious and other Metals for Enameling* H de Konigh Henley Publishing Company, New York

*Designs in Metal* Paul Bridge and Austin Crossland Batsford, London

*Metal: Design and Technique* Wilhelm Braun-Feldweg Batsford, London: Van Nostrand Reinhold, New York

*Introducing Jewellery Making* John Crawford Batsford, London: Watson-Guptill, New York

*The Technique of Jewellery* Rod Edwards Batsford, London

# Suppliers
## Great Britain

### General (stocking a wide range of enamelling requisites)

The Enamel Shop
  21 Macklin Street
  London, WC2B 5NH

Dryad
  Northgates
  Leicester, LE1 9BU

### Blow torches

Charles Cooper (Hatton Garden) Ltd
  Knights House
  23-27 Hatton Wall
  London, EC1 8JJ

Frank Pike Ltd
  15 Hatton Wall
  Hatton Garden
  London, EC1N 8JE

H. S. Walsh & Sons Ltd
  12-16 Clerkenwell Road
  London, EC1M 5PL
  *and*
  243 Beckenham Road
  Beckenham, Kent, BR3 4TS

### Brushes, glass fibre

Charles Cooper (Hatton Garden) Ltd
  Knights House
  23-27 Hatton Wall
  London, EC1 8JJ

H. S. Walsh & Sons Ltd
  12-16 Clerkenwell Road
  London, EC1M 5PL
  *and*
  243 Beckenham Road
  Beckenham, Kent, BR3 4TS

### Carborundum

Hardware stores

### Copper sheet

J. Smith & Sons (Clerkenwell) Ltd
  St John's Square
  London, EC1P 1ER

### Crucibles

Charles Cooper (Hatton Garden) Ltd
  Knights House
  23-27 Hatton Wall
  London, EC1 8JJ

Morganite Refractories Ltd
  Neston
  South Wirral
  Cheshire, L64 3RE

Frank Pike Ltd
  15 Hatton Wall
  Hatton Garden
  London, EC1N 8JE

H. S. Walsh & Sons Ltd
  12-16 Clerkenwell Road
  London, EC1M 5PL
  *and*
  243 Beckenham Road
  Beckenham, Kent, BR3 4TS

## Enamels

W. G. Ball Ltd
  Anchor Road
  Longton
  Stoke-on-Trent, ST3 1JW

Ferro (Great Britain) Ltd
  Wombourne
  Wolverhampton, WV5 8DA

## Kilns, enamelling

W. G. Ball Ltd
  Anchor Road
  Longton
  Stoke-on-Trent, ST3 1JW

R. M. Catterson-Smith Ltd
  Woodrolfe Road
  Tollesbury
  Maldon
  Essex, CM9 8SJ

Ferro (Great Britain) Ltd
  Wombourne
  Wolverhampton, WV5 8DA

Kilns & Furnaces Ltd
  Keele Street
  Tunstall
  Stoke-on-Trent, ST6 5AS

## Kilns, pottery, front-loading

Bricesco Combustion Services Ltd
  Bricesco House
  Park Avenue
  Wolstanton
  Newcastle
  Staffordshire, ST5 8AT

R. M. Catterson-Smith Ltd
  Woodrolfe Road
  Tollesbury
  Maldon
  Essex, CM9 8SJ

Ferro (Great Britain) Ltd
  Wombourne
  Wolverhampton, WV5 8DA

The Fulham Pottery Ltd
  184 New Kings Road
  London, SW6 4PB

Kilns & Furnaces Ltd
  Keele Street
  Tunstall
  Stoke-on-Trent, ST6 5As

## Lacquer for protecting polished copper

W. Canning Materials Ltd
  Great Hampton Street
  Birmingham, B18 6AS

## Lustres

The Enamel Shop
  21 Macklin Street
  London, WC2B 5NH

## Masks, protective

Smog masks from pharmacies

## Metalworking tools

*Simple tools for metalwork*

Hardware stores and other tool suppliers

*Special tools*

Charles Cooper (Hatton Garden) Ltd
  Knights House
  23-27 Hatton Wall
  London, EC1 8JJ

Frank Pike Ltd
  15 Hatton Wall
  Hatton Garden
  London, EC1N 8JE

H. S. Walsh & Sons Ltd
  12-16 Clerkenwell Road
  London, EC1M 5PL

*and*
  243 Beckenham Road
  Beckenham, Kent, BR3 4TS

## Nitric acid

Obtainable from some pharmacies

## Polishing motors, mops and materials

W. Canning Materials Ltd
  Great Hampton Street
  Birmingham, B18 6AS

Charles Cooper (Hatton Garden) Ltd
  Knights House
  23-27 Hatton Wall
  London, EC1 8JJ

Frank Pike Ltd
  15 Hatton Wall
  Hatton Garden
  London, EC1N 8JE

H. S. Walsh & Sons Ltd
  12-16 Clerkenwell Road
  London, EC1M 5PL
  *and*
  243 Beckenham Road
  Beckenham, Kent, BR3 4TS

## Sprayers, electric (for spraying gum solution)

Ferro (Great Britain) Ltd
  Wombourne
  Wolverhampton, WV5 8DA

**Stencil knives**

Hobby and artists' materials stores

**Tongs, copper and brass**

Frank Pike Ltd
  15 Hatton Wall
  Hatton Garden
  London, EC1N 8JE

H. S. Walsh & Sons Ltd
  12-16 Clerkenwell Road
  London, EC1M 5PL
  *and*
  243 Beckenham Road
  Beckenham, Kent, BR3 4TS

**Tweezers, pointed**

Jewellers' requisites suppliers

Charles Cooper (Hatton Garden) Ltd
  Knights House
  23-27 Hatton Wall
  London, EC1 8JJ

Frank Pike Ltd
  15 Hatton Wall
  Hatton Garden
  London, EC1N 8JE

H. S. Walsh & Sons Ltd
  12-16 Clerkenwell Road
  London, EC1M 5PL
  *and*
  243 Beckenham Road
  Beckenham, Kent, BR3 4TS

# Suppliers

## U.S.A.

### Enamels, kilns, basic equipment and supplies

Allcraft Tool and Supply Co., Inc.
  100 Frank Road
  Hicksville
  NY 11801

American Art Clay Co.
  4717 West 16th Street
  Indianapolis
  IN 46222

Sax Arts and Crafts
  P.O.B. 2002
  Milwaukee
  WI 53201

Thompson Enamel
  P.O.B. 310
  Newport
  KY 41072

### Special sources

*(wire-brushed copper panels and shapes, special tools, racks, enamels and other supplies)*

Seaire
  17909 S. Hobart Blvd
  Gardena
  Calif. 90248

### Kilns only

Cress Manufacturing Co., Inc.
  1718 Floradale Avenue
  S. El Monte
  Calif. 91733

### Miscellaneous supplies

*Abrasives (carborundum, emery paper and pumice)*
Hardware stores

*Acid*
Chemical suppliers, pharmacies and enamelling suppliers

*Copper*
Sheet: industrial metal suppliers
Sheet, shapes and panels: enamelling suppliers
Tooling foil: metal suppliers, craft and hobby stores

*Lacquers and waxes*
Hardware stores

*Propane hand torches and asbestos products*
Hardware stores

# Index

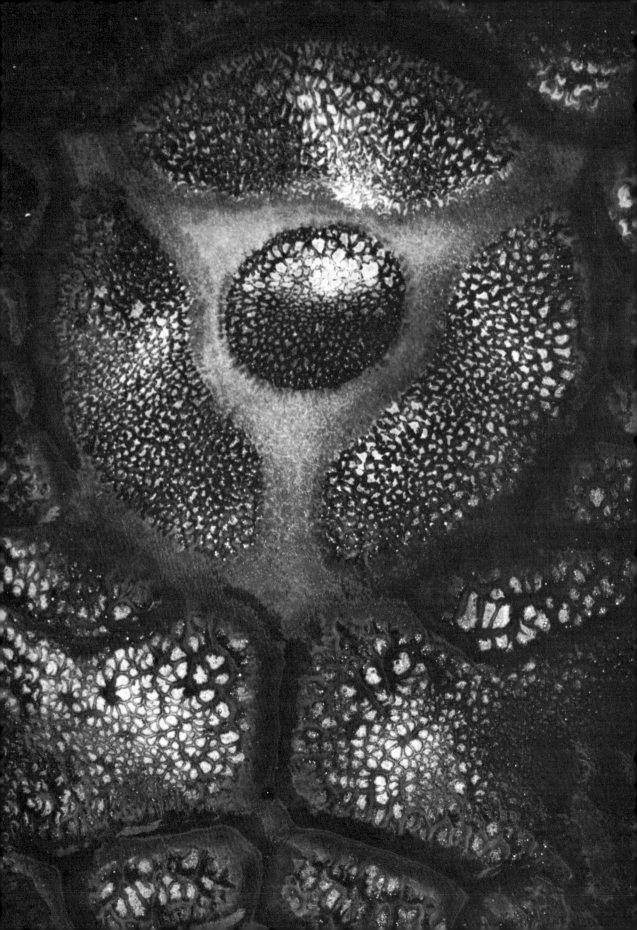

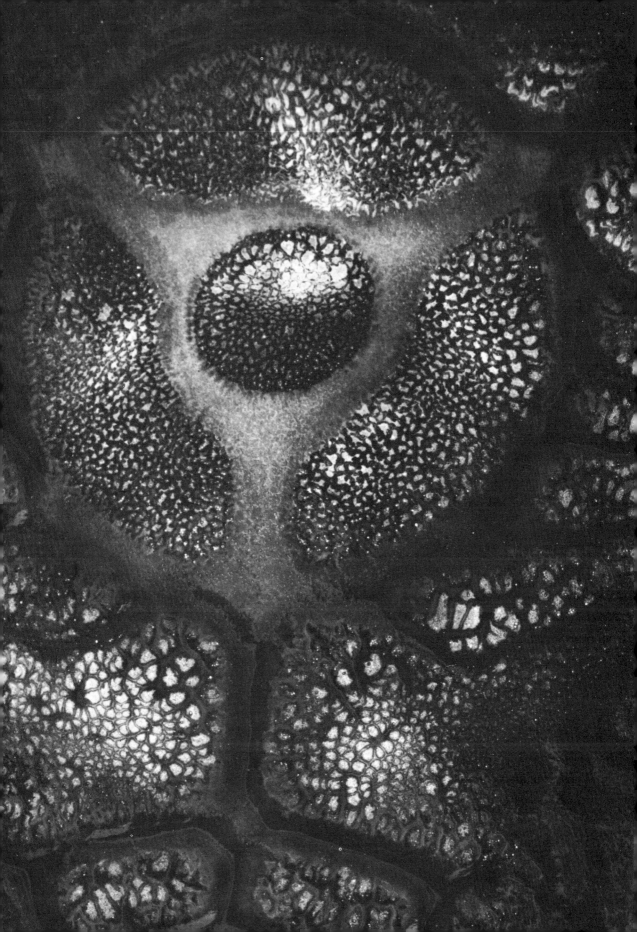